Sparse Matrix
Proceedings
1978

# Sparse Matrix Proceedings 1978

editors: **Iain S. Duff**
*AERE Harwell*

and

**G. W. Stewart**
*University of Maryland*

Philadelphia

Library of Congress Catalog Card Number: 79-88001

ISBN: 0-89871-160-6

# Table of Contents

# Introduction

The papers in this book were presented at the Symposium
on Sparse Matrix Computations held in the Hyatt Regency,
Knoxville, Tennessee on November 2-3, 1978. The con-
ference was organized as a SIAM activity under the Chair-
manship of Donald Rose by a programming committee con-
sisting of James R. Bunch, Alan George, Robert J. Plemmons,
John K. Reid, G. W. Stewart, Robert E. Tarjan, Richard S.
Varga, and Robert C. Ward. Support was provided by the
Army Research Office, the Office of Naval Research, and
Oak Ridge National Laboratory.

A major aim of the committee was to emphasize applications-oriented research. Four speakers were invited to
present papers in different application areas and the
rest of the papers were selected by the program committee
from extended abstracts submitted by potential speakers.
This enabled the elimination of some obvious circle
squarers and angle trisectors hopefully without preventing
the early dissemination of new results. To expedite pub-
lication of the proceedings the authors were asked to
prepare their papers as camera ready scripts, and to take
full responsibility for their content. However, we take
responsibility for the following comments, which are intended to serve the reader as an introductory guide to
the papers.

At one stage we thought of arranging the papers under
a few headings such as applications, software, and algorithms, but it was not long before we discovered that
most contributions fell into two or more such sets.

Indeed we feel it is one of the more pleasant aspects of
this field that a single paper can have a foot in several
camps and still be valuable and coherent. We therefore
have chosen to arrange the papers in the order in which
they were presented at the symposium.

Before describing the contributions in a little more
detail it is worthwhile reflecting that the origins of
our subject go back further than the ten years spanning
the conferences on sparse systems. According to Wilson,
at this conference, "the substructure concept has been a
fundamental approach to the analysis of complex structural
systems for over twenty years", and Tinney and his colleagues were running sparse ordering algorithms, including
minimum degree, some years before the first meeting dealing explicitly with sparse matrices and their applications.

It is just over ten years since the first conference
on sparse matrices and their applications was held at
IBM Yorktown Heights, New York, and it is instructive to
glance back at the faded proceedings to see how much
sparse matrix technology has changed over the decade.
The composition of attendees was similar to that of the
Knoxville meeting, with a very high percentage from
industry who were interested in the application of sparse
matrix algorithms and software to real problems. Indeed,
an active group of people oriented towards applications
has characterized this branch of numerical mathematics
and kept the subject from degenerating into an esoteric
backwater of jargon and theorems proved for their own
sake. The proceedings of this conference, as those of
the three intervening conferences, have reflected this
with a wealth of papers about applications. The areas
most strongly represented by contributions at this conference are structural analysis, chemical engineering,
image processing, and linear programming. We see in
these contributions that applications do not merely siphon
off and apply ideas but actively contribute to research

with new ideas from other specialities. It is this inter-
play that makes the study of sparse systems such an exciting area to work in. We hope that the reader will
share some of this excitement with us.

We have stressed this particular theme because we
believe it to be of vital importance, but other threads
also span the decade. We see the debate between direct
and iterative methods still around, with new semi-direct
techniques blunting the distinction and adding a new
dimension. This particular debate now appears more
informed and less impassioned than formerly, with prac-
titioners of the various approaches more aware of the
limitations of their techniques. Another common theme,
which permeates all five proceedings, is the use of graph
theory to model sparse matrices and the operations on
them. Undoubtedly, this is a very useful tool, although
it is refreshing to see that the use of such techniques
has definitely matured, and there is less tendency to
employ obscure graphical formalism than earlier. The
efficient use of data structures and I/O are as important
considerations today as they were at Yorktown, and so
is the influence of new and forthcoming machine architectures on the design of software.

At Yorktown in 1968 it was evident that the subject
was in its infancy and contributors were trying to define
goals and identify difficulties. Now the walls of the
playpen are down and we can have some idea of our
crawling speed by looking at the concerns of the panel
at the end of the 1968 meeting. We feel that there is
now a much greater awareness of the potential and limitations of advances in computing systems. We are still
worried about the difficulties of data transfer and the
ease with which otherwise efficient algorithms can be
hamstrung by I/O overheads. However, a body of practical
experience in this area has turned these difficulties
into well-defined problems, rather than ogres lurking in

shadows. Admittedly the only thing we have learnt about scaling during the past few years has been to establish the infeasibility of a general all-purpose algorithm. It is now clear that the scaling process should be separated from the rest of the computation and that the ideal scaling must really be obtained from information provided by the user. It is gratifying that our ability to handle larger and larger problems is as much due to the improvement of algorithms and software as to the increase in computing power of modern machines. The upper limit on the order of system being solved was stated by Willoughby in 1968 to be $10^4$, whereas we see today that, in some application areas, systems of order $10^5$ are commonly encountered and handled successfully. Another area in which great advances have been made is in the solution of very large eigensystem problems. To our knowledge, this is the first conference where a practical method has been described for obtaining all the eigenvalues of a very large matrix, thus answering in the affirmative the question posed by members of the audience some ten years ago.

We now turn to the discussion of the individual papers and their relation to sparse matrix technology in general. A number of papers describe techniques and software for the solution of linear systems in which the coefficient matrix is symmetric and positive definite. It is worth noting that one of the most influential papers in this area, by Speelpenning, has never been published. Two different interpretations and applications of his generalized element approach are presented here, one by George and Liu and the other by Eisenstat et al. George and Liu make heavy use of what Karp calls a legal path, observing that, if vertices i and j are connected by a path which only contains vertices corresponding to variables eliminated by Gaussian elimination before i and j, then they will be joined by an edge in the reduced graph which, of

course, corresponds to a nonzero in position (i,j) of the
LDL$^T$ factors. They implement this notion by the use of
supernodes, which are sets of vertices corresponding to
eliminated variables that are adjacent in the reduced
graph. They use this implementation to develop an
algorithm that can perform a minimum degree ordering,
often in time proportional to the number of variables
and in an amount of storage which is predetermined and
independent of the numbers of nonzero elements in the
factors. Eisenstat et al consider the superelements
composed of all vertices connected to such supernodes
and represent the factorization as a sequence of element
amalgamations, followed by elimination of vertices
interior to the new elements. They assume that an ordering
of the variables for elimination already exists and con-
centrate on efficient methods for organizing the creation
of the numerical factors. Their algorithm can be tuned
to the amount of store available and for typical problems
will run in about one quarter of the storage at a cost
of about 2.5 the normal computation, owing to the re-
computation of intermediate factors that have been discarded to achieve the storage gains. The algorithm of
Manteuffel also considers a method for handling symmetric
positive definite systems in which he performs ICCG on
the matrix

$$I + \frac{1}{1 + \alpha} \, B$$

where the coefficient matrix A has first had its diagonal
scaled to unity and then has its off diagonals reduced
by the factor $(1+\alpha)^{-1}$. He illustrates the effect of the
choice of $\alpha$ by experiment, showing that substantial
gains can be obtained and that the use of this shifted
incomplete Cholesky can extend the range of matrices
to which ICCG may be successfully applied.

   Duff examines the performance of techniques similar
to the two discussed in the previous paragraph, showing
the great gains which can be realized on certain classes

of problems. However, the main part of his paper is concerned with a practical comparison of alternative devices employed by codes for the in-core solution of general unsymmetric systems. It is seen that compiled or interpretative code approaches are competitive when the matrix is very sparse, that some form of numerical control (for example, threshold pivoting) is desirable, and that significant gains can be obtained by going over to full matrix code towards the end of the elimination. The feasibility of performing a preordering to block triangular form is also established.

The area of linear programming has stimulated research in sparse techniques for many years. This conference has three papers devoted to this topic. Gay shows the feasibility of combining two schemes discussed in earlier literature. If one holds the upper triangular part of the factors of the basis matrix in the form

$$\begin{pmatrix} I & R \\ 0 & F \end{pmatrix}$$

where F is a small square relatively dense submatrix, then we can maintain the basis factors during subsequent updates by keeping only F in rapid random access memory since additional spike columns can be written onto the end of R, which is held in auxiliary store. He combines this scheme with one that replaces numerical elimination operations as much as possible with permutations when performing the update on F and illustrates the practical benefits of such a scheme. Partitioning methods for the factorization of the initial basis are also considered and those, like $P^4$, which reduce the number of spike columns (order of F) are preferred. Dantzig and Perold examine another block triangular method particularly applicable to time-staged and multi-staged linear programs, where it can be shown that groups of variables persist in the basis over several consecutive time periods.

Updating schemes for such a block triangular basis factorization are discussed in some detail and computational experience on several dynamic models is reported. In the paper of O'Leary two different techniques are combined to form a working algorithm. She considers linear programming problems arising in connection with partial differential equations. These are often characterized by matrices with a block band structure in which the blocks themselves are sparse. She uses the active set strategy of Gill and Murray, which gives rise to a subproblem involving the solution of normal equations, which is computed by a conjugate gradient algorithm preconditioned by SSOR. The use of the method is illustrated by a problem arising from a pumping model in drainage.

The paper by Bank and Sherman describes a complete system for the solution of boundary value problems in partial differential equations. Techniques for storing and refining the triangulation of a user-defined initial triangle are described and these are used with an adaptive multi-grid approach to produce a finite element solution to the original problem. They discuss in some detail the user interface and describe results of their adaptive and nonadaptive schemes on a model problem with a crack singularity.

We have mentioned the eigensystem paper of Cullum and Willoughby in our introductory remarks. They carry the Lanczos recurrence beyond the order n of the matrix A to generate a tridiagonal "projection" $T_m$, of order $m \geq n$. Spurious eigenvalues of $T_m$ (that is, eigenvalues which are not eigenvalues of the original matrix) are identified by examining the eigenvalues of the sub-matrix obtained by removing the first row and column of $T_m$. They find experimentally that they can identify nearly all of the isolated eigenvalues of A with $m = 2n$ although to obtain them all may require up to 6n recurrences.

We have already noted at some length the healthy

interaction between sparse matrix research and applications, and many of the papers just discussed illustrate their methods by applying them to practical problems. We now discuss some papers concerned explicitly with applications. Wilson reviews techniques which are used in the solution of large systems (order $10^4$ and greater) from problems in structural analysis. He stresses that repetition in structures can often be used to great advantage. He also compares the methods of frontal solution, where assembly and elimination are concurrent, with out of core blocked profile elimination. The latter appears better suited to extremely large structures while the frontal scheme is much faster when the active matrix can be held in core. He also considers combining iterative and direct methods with a partitioning of the original system in the form

$$
\begin{pmatrix} K_{rr} & K_{rg} \\ K_{gr} & K_{gg} \end{pmatrix}
\begin{pmatrix} \Delta \\ u_g \end{pmatrix}
=
\begin{pmatrix} P_r \\ P_g \end{pmatrix}
$$

where iterative schemes will work well on systems with the coefficient matrix $K_{rr}$ while it is feasible and preferable to use a direct solver on $K_{gg}$, which is typically twenty times smaller than the main system. The use of sparse techniques in the optimization of large chemical processes is discussed by Westerberg and Berna. Again they encounter systems of order $10^5$ in quadratic programming problems arising from the linearization of the constraints and Lagrangian derivative of a general optimization problem. Their method is designed to solve systems, in which the matrix has a loosely connected block diagonal structure, the factorization being performed separately on submatrices of the original system with pivoting restricted to specified rows and columns in order to avoid fill-in in zero blocks of the original structure. They also describe a high level language,

LASCALA, which they have designed to control the input
and solution of such systems.  The subject of image
reconstruction from projections is the topic of the paper
by Herman.  His examples are drawn primarily from X-ray
projections of human or animal bodies and the linear
systems which result can easily be of order $10^5$.  Often,
particularly in the medical field, it is important to
obtain a fast solution to these problems with a relatively
low level of accuracy.  He discusses in some detail the
application of the Bayesian approach to image reconstruction and describes his use of Richardson's iterative
scheme to solve the resulting system.  Alternative
solution techniques, grouped loosely under the term row
generation methods, are explored in detail in the paper
by Censor and Herman.  This paper admirably illustrates
how general purpose methods can spring from research in
a particular application.  Row generation methods are
defined as any iterative procedure which, without making
any changes to the original matrix A, uses the rows of
A a row at a time.  As the authors remark, they are
effectively working with systems where there is no matrix
at all -- all the information required for solution is
generated equation by equation.  Row generation methods
include such well known algorithms as the projection
method by Kaczmarz, and Censor and Herman continue to
describe and classify a whole battery of such techniques,
in each case giving a very clear geometric interpretation
of the method.

Although the beating heart analogy that introduces
the Kung and Leiserson paper may cause one to conclude
that it deals with another medical application, the paper
in fact represents an attempt to apply new concepts of
computer architecture to matrix computations.  They show
how it is possible to design arrays of processors to
compute matrix by vector products, matrix by matrix
products, L-U decompositions, and solutions of tri-

angular systems. The idea of a special purpose machine
to solve our linear equations is very attractive particularly if we consider that the procedure can be
completed in $O(n)$ time, where n is the order of the system.
The simplicity of their processors and the rapidly ex-
panding integrated circuit technology give a genuine
hope for the practical realization of their suggestions.

Iain S. Duff
*AERE Harwell, England*

G.W. Stewart
*College Park, Maryland*

# Solution of Sparse Stiffness Matrices
# for Structural Systems

**Edward L. Wilson***

Abstract. The equilibrium equations, which govern the
behavior of structural systems, may have very special properties. In most cases, repeated parts of the structure
cause the generation of identical terms in the stiffness
matrix. If these properties are recognized a significant
reduction in both the number of numerical operations and
the required computer storage may be realized.

Both the frontal and the active column methods are special cases of Gauss elimination and have been used extensively in the solution of structural systems. While the
frontal approach eliminates all operations on nonzero
terms, the active column storage scheme has certain practical advantages because it has unlimited capacity on small
computers and can be extended to the extraction of eigenvalues and eigenvectors.

Iteration methods have not proven effective for most
two-dimensional structures as compared to a direct solution
approach. However, for three-dimensional systems which
generate large "band width" matrices, iterative methods
coupled with a coarse mesh direct solution have great
potential. In addition, iterative methods may be advantageous for the solution of finite element systems which are
solved by an adaptive mesh refinement approach.

1. Introduction. The purpose of this paper is to

present a summary of several different numerical methods

for the solution of linear structural systems subjected to

static loading. The presentation will emphasize the physi-

cal behavior of different structures as opposed to the

mathematical properties of the governing equations. It will

be illustrated that the best method will depend on the type

of structural system and that each method can be optimal

for a specific problem. In addition, a combination of

*Department of Civil Engineering, University of California,
  Berkeley, California.

1

methods is often ideal.

A schematic representation of a large offshore drilling platform is shown in figure 1. One notes the symmetry and repeated geometry of the structural components. Most other types of structures have similar repeated geo- metric properties in order to minimize production and fabrication costs.

For the purpose of analysis (the evaluation of joint dis- placement and member forces) of a structure as shown in figure 1, it is necessary to transform the physical struc- ture into a mathematical model. This model, in most cases, has a finite number of nodes or joints.

The offshore structure shown can be idealized by one- dimensional structural elements each connected to two nodes.

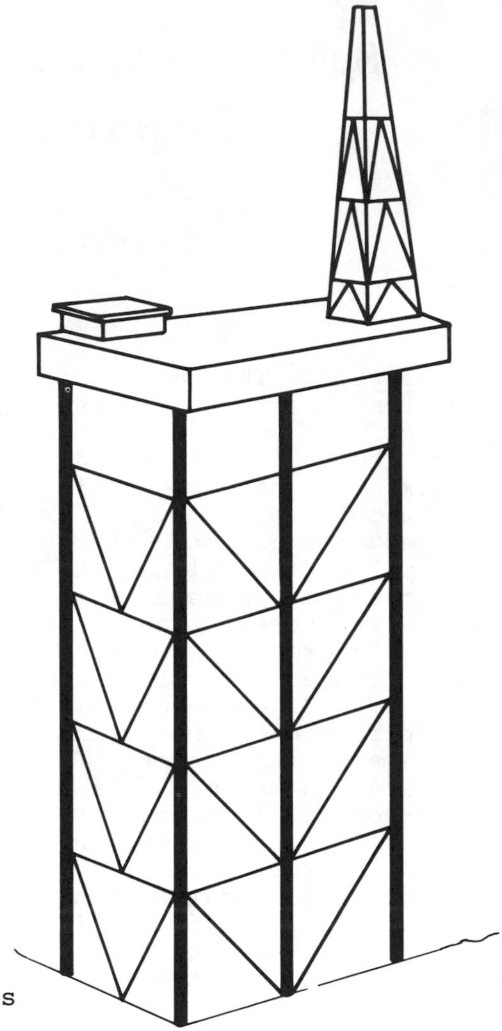

FIGURE 1   Schematic of Large
           Offshore Structure

Other types of structures can be idealized by other types of elements.  Figure 2 shows several different types of struc- tural elements commonly used in the idealization of struc- tural systems.  The basic equations for each element express node forces in terms of node displacements.  Or

$$F^{(m)} = K^{(m)} U \tag{1}$$

where $F^{(m)}$ is the node force vector associated with element "m" and $K^{(m)}$ is the corresponding element stiffness matrix. The vector U contains the node displacements which are

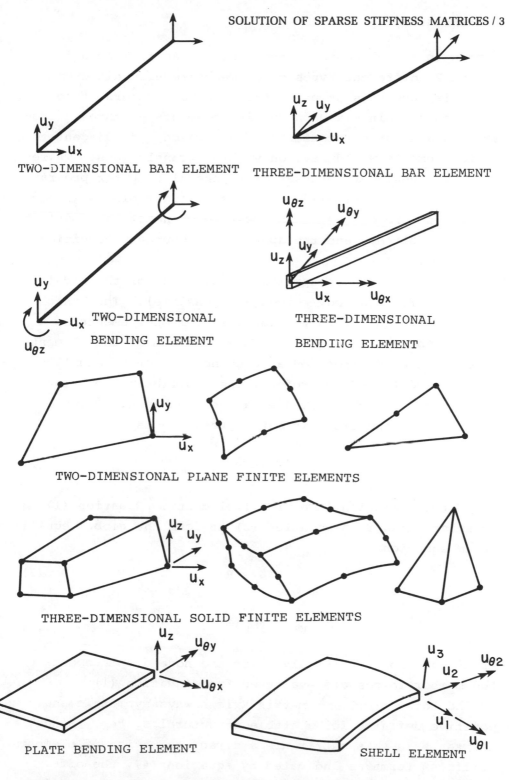

TWO-DIMENSIONAL BAR ELEMENT    THREE-DIMENSIONAL BAR ELEMENT

TWO-DIMENSIONAL
BENDING ELEMENT

THREE-DIMENSIONAL
BENDING ELEMENT

TWO-DIMENSIONAL PLANE FINITE ELEMENTS

THREE-DIMENSIONAL SOLID FINITE ELEMENTS

PLATE BENDING ELEMENT                SHELL ELEMENT

FIGURE 2   Summary of Different Element Types

common to all elements in the system. As indicated in
figure 2, different types of elements require different
node displacements as unknowns. The most general type of
node can contain six unknown displacements -- three translations and three rotations. Nodes which are adjacent to
solid elements would have only three translations. Therefore, the engineer, or user of a general computer program
for structural analysis, must have considerable knowledge
of element properties if an accurate mathematical model is
to be established and the appropriate boundary conditions
specified.

External loads are applied at the nodes of the model
(also generalized body forces are possible). The basic
equations which govern the total system of elements is a
simple statement of equilibrium -- the sum of the element
forces (in each direction at each node) must be equal to
the applied loads. Therefore, one equilibrium equation is
developed for each nodal displacement unknown. Or, the
equilibrium conditions in matrix notation is

$$\sum_m F^{(m)} = \underline{R} \tag{2}$$

where the summation is over all elements. Equation (1)
is substituted into equation (2) to form the global equilibrium equations

$$K \ U = R \tag{3}$$

where

$$K = \sum_m K^{(m)} \tag{4}$$

After equation (3) is solved for the node displacements $\underline{U}$
the member forces are evalauted from equation (1).

The total stiffness matrix $\underline{K}$ is always symmetric and
positive definite for a stable structural system. Since
the terms in this matrix are a direct summation of element
stiffness terms as indicated by equation (4), the off-
diagonal terms are related to the connectivity of nodal

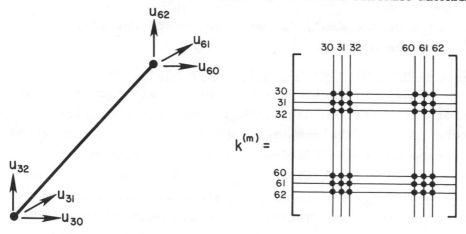

FIGURE 3   Typical Element Stiffness Matrix

unknowns by each element in the system.  As an example,
the bar element shown in figure 3 would cause 36 terms to
exist in the global stiffness matrix corresponding to the
row and column numbers indicated by the six node displacements.  Therefore, one can determine the location of terms
in the stiffness matrix by a direct examination of the
mathematical model and the numbering system for the nodal
displacements.

It is apparent that the complete development and use of
a general computer program for structural analysis requires
background in four areas.  First, in order to develop
accurate element properties it is necessary to have a
stong background in structural mechanics.  Second, the
efficient formation of element stiffness matrices and solution of the global equilibrium equations requires the use
of appropriate numerical techniques.  Third, these methods
must be effectively implemented on a computer system.
Fourth, the use of the program for the analysis of a continuous structure requires an understanding of structural
behavior in order that the appropriate elements and mesh
are used in the selection of the mathematical model which
is solved by the computer program.  The purpose of this
paper is to review numerical methods which are used in the
solution of equilibrium equations.  For many structures

this phase of the analysis can demonstrate the efficiency
of a computer program.

   2.   Direct Solution of Equilibrium Equations.   All
practical direct solution methods are versions of elimination or factorization algorithms.   The most significant
difference is in the computer implementation and the method
of storage of the coefficients in the stiffness matrix.
Systems of 10,000 equations with 10,000,000 nonzero terms
in the triangularized stiffness matrix are not uncommon for
structural problems.   Therefore, incore methods cannot be
considered as a general approach.   At the present time the
factorization of the stiffness matrix stored in blocked
profile form or the direct Gauss elimination of the equilibrium equations as they are formed (frontal method) appear
to be the most effective approaches.

   Profile solvers are well documented in references [1]
to [5].   Prior to solution, the equilibrium equations are
formed in blocks of columns approximately equal to one-half
of the available high-speed core storage and all blocks are
placed in secondary storage.   The factorization in sequential block order is performed by reading the required previously factored blocks into high-speed storage for each
block to be factored.   Therefore, all accesses to secondary
storage are for a large block of data.   In addition, the
number of numerical operations between reads is significant
since one large block of data is operating on another large
block of data.   Therefore, the access time is small compared
to the incore numerical effort for most structural problems.

   The frontal approach solves the set of equilibrium
equations as they are formed [6].   The summation given by
equation (4) is performed as element stiffness matrices are
formed.   When all element stiffness matrices associated with
nodes are complete the unknowns associated with that node
are eliminated and the current stiffness matrix is compacted
and the block substitution equations are placed on low-
speed storage.

At any step of the method only the stiffness coefficients associated with the nodes with incomplete stiffnesses are retained in high-speed storage. This set of nodes forms the "front" and governs the capacity of the method since the stiffness coefficients associated with the unknowns must be retained in high-speed core storage if the method is to retain its maximum efficiency.

There are advantages and disadvantages associated with both the profile (or envelope) and frontal (or wave front) methods. Table 1 summarizes a comparison of the methods. It is clear that for problems where the "front" can be contained in core the frontal method will be the same or faster than profile solvers. However, for very large problems the blocked profile approach may be the only practical solution method.

3. Substructure Methods. The substructure concept has been a fundamental approach to the analysis of complex structural systems for over twenty years. However, it has not been used extensively in general purpose programs since it does not offer significant numerical benefits unless parts of the structures are identical. In order to illustrate the method the finite element system shown in figure 4 is divided into four identical substructures. The governing equations for these parts of the system can be written in matrix partition form as

$$\begin{bmatrix} K_{ii} & K_{ir} \\ K_{ri} & K_{rr} \end{bmatrix} \begin{bmatrix} U_i \\ U_r \end{bmatrix} = \begin{bmatrix} F_i \\ F_r \end{bmatrix} \tag{5}$$

where $U_i$ are the displacements of the interior nodes which are to be eliminated and $U_r$ are the displacements at the boundary nodes and are to be retained as unknowns. Solving for $U_i$

$$U_i = K_{ii}^{-1}[F_i - K_{ir}U_r] \tag{6}$$

TABLE 1  Comparison of Profile and Frontal Methods

|  | PROFILE | FRONTAL |
|---|---|---|
| Capacity | Governed by amount of low-speed computer storage | Stiffness of "front" should be retained in core for maximum effectiveness |
| Speed | Good for both large and small problems | Excellent for small to medium problems; poor if front must be partitioned |
| Additional load vectors after factorization | No problem | Slow compared to profile |
| Programming problems | Equation solver is a separate subroutine | Equation solver is often embedded into stiffness formulation subroutine |
| Time required to obtain global stiffness matrix | Can be slow for a large number of elements or blocks | Very fast since element search is not required |
| Operation on zero coefficients | Possible for certain systems | Not possible for all systems |
| Logical operations during solution | Very small in all cases | Can be significant |
| Extension to dynamics | No problem | Slow due to additional logical operators |
| Extension to substructure analysis | No problem | No problem |
| Ability to exploit speed of vector or parallel computers | Excellent since all operations are on long columns | Some increase in speed is possible |
| User knowledge required | None, if profile minimization is used | User normally specifies order of elimination by element ordering |

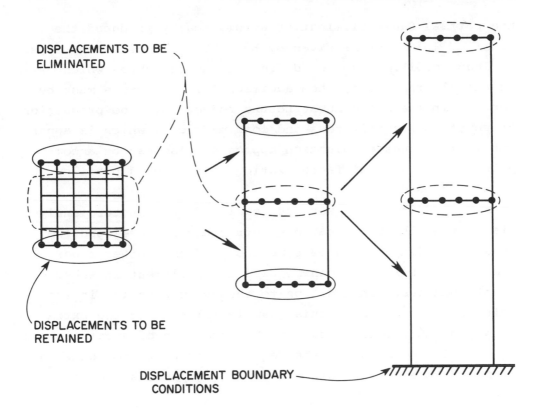

DISPLACEMENTS TO BE
ELIMINATED

DISPLACEMENTS TO BE
RETAINED

DISPLACEMENT BOUNDARY
CONDITIONS

FIGURE 4   Application of the Substructure Method

which leads to

$$\tilde{K}_r \, U_r \;\; = \;\; \tilde{R}_r \tag{7}$$

in which

$$\tilde{K}_r \;\; = \;\; K_{rr} - K_{ri} \, K_{ii}^{-1} \, K_{ir} \tag{8a}$$

$$\tilde{R}_r \;\; = \;\; R_r - K_{ri} \, K_{ii}^{-1} \, R_i \tag{8b}$$

The stiffness of the substructure $\tilde{K}_r$ is with respect to
the displacements of the fourteen retained nodes and has the
normal physical properties of force per unit of deformation
with respect to the retained nodes.  The substructure stiffness matrix, given by Eq. (8a), is not normally developed by
the matrix operations indicated.  The same submatrix, $\tilde{K}_r$,
is developed by the direct application to Eq. (5) of the
Gauss elimination algorithm to the unknowns $U_i$.  In addition,

the partial Gauss elimination automatically produced the modified load vectors given by Eq. (8b). If different types of load conditions act on different substructures which have identical properties, then additional load vectors must be carried in the basic substructure reduction. The production of substructure stiffness and load matrices, which is apparent in the Gauss elimination approach, also can be accomplished by a partial factorization if a blocked profile method is used [5].

Figure 4 indicates various levels of substructure analysis. After the first basic substructure is developed two such superelements can be combined to form another superelement which in turn can be used as an element in subsequent analyses. In addition to the obvious saving in the solution of equations, this example illustrates the extra saving of numerical effort in the formation of equations and an overall reduction in the required computer storage when compared to a direct solution of the complete system of elements.

4. A Frontal Substructure Method for the Three-Dimensional Analysis of Complex Building Systems. The analysis of building systems can involve a solution of a very large number of equations. For example, a 50-story building with 50 vertical column lines would generate 150,000 equations with a band width of 312 and would cost over $5000 at current service bureau rate. However, if appropriate numerical methods are used for this class of problem the computer costs can be less than $50 [7].

Figure 5 illustrates the important structural elements of a small building system. Most buildings are designed so that the in-plane stiffness of the floor systems are very stiff compared to the stiffness properties of the individual frame substructures. Therefore, the independent displacements for this type of system are two translations and one rotation of each floor slab and two rotations and one axial displacement for each column as illustrated in figure 6.

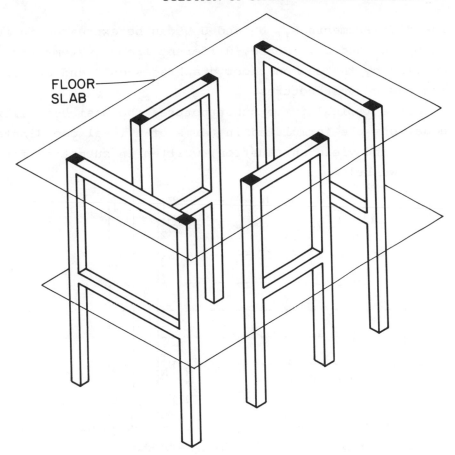

FIGURE 5   Schematic of Typical Building System

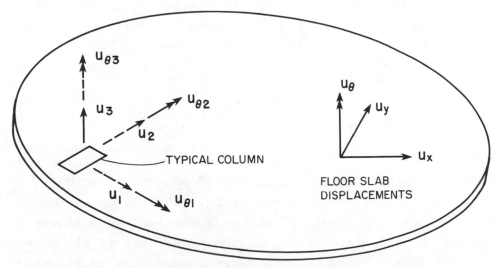

FIGURE 6   Unknown Displacements for Building System

The displacements $U_1$, $U_2$ and $U_{\theta 3}$ can be expressed in terms of $U_x$, $U_y$ and $U_\theta$. Hence, different frame systems are coupled by the common floor diaphragms only and are used as the basic substructures.

For a typical frame substructure, the basic equilibrium equations are formulated in terms of a local coordinate system and yield an equation, written in submatrix form, as shown below.

$$
\begin{bmatrix}
F_1 \\
F_2 \\
F_3 \\
-- \\
-- \\
-- \\
F_N \\
F_L
\end{bmatrix}
=
\begin{bmatrix}
K_1 & C_1 & 0 & 0 & -- & -- & -- & E_1 \\
C_1^T & K_2 & C_2 & 0 & -- & -- & -- & E_2 \\
0 & C_2^T & K_3 & C_3 & -- & -- & -- & E_3 \\
 & 0 & 0 & -- & -- & -- & -- & -- & -- \\
-- & -- & -- & -- & -- & -- & -- & -- \\
-- & -- & -- & -- & -- & -- & -- & -- \\
-- & -- & -- & -- & -- & -- & K_N & E_N^T \\
E_1^T & E_2^T & E_3^T & -- & -- & -- & E_N^T & K_L
\end{bmatrix}
\begin{bmatrix}
U_1 \\
U_2 \\
U_3 \\
-- \\
-- \\
-- \\
U_N \\
K_L
\end{bmatrix}
\tag{9}
$$

This equation is solved by the elimination of the unknowns $U_1$ to $U_N$ by a frontal solution method.

The assembly of Eq. (9) and frontal reudction is carried out systematically story level by story level. Furthermore, only a small fraction of Eq. (9) need be retained in high-speed computer storage as indicated below.

$$
\begin{bmatrix}
\bar{\bar{F}}_n \\
\bar{F}_{n+1} \\
F_L + \bar{F}_L
\end{bmatrix}
=
\begin{bmatrix}
\tilde{K}_n & \tilde{C}_n & \tilde{E}_n \\
\tilde{C}_n^T & \tilde{K}_{n+1} & \tilde{E}_{n+1} \\
\tilde{E}_n^T & \tilde{E}_{n+1} & \tilde{K}_L
\end{bmatrix}
\begin{bmatrix}
U_n \\
U_{n+1} \\
U_L
\end{bmatrix}
\tag{10}
$$

where $\bar{F}$, $\tilde{K}$, $\tilde{C}$, $\tilde{E}$ and $\tilde{K}_L$ indicate terms which have been modified due to the Gauss elimination of unknown $U_1$ to $U_{n-1}$. With the storage scheme given by Eq. (10) the operation required at each level "n" may be summarized as follows:

1. Add to the frame stiffness the properties of all members between levels n and n+1 which are shown below in figure 7.

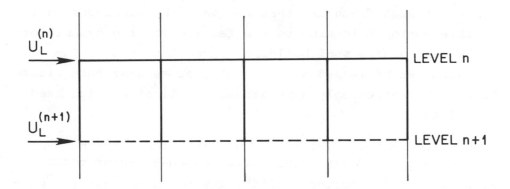

FIGURE 7    Formation and Reduction of Frame Stiffness

2.    Eliminate the unknowns $U_n$ from Eq. (10).

3.    Save the backsubstitution equations associated with level n.

4.    Rearrange the terms in storage in order to modify Eq. Eq. (10) to the following form:

$$\begin{bmatrix} \bar{F}_{n+1} \\ 0 \\ F_L + \bar{F}_L \end{bmatrix} = \begin{bmatrix} \tilde{K}_{n+1} & 0 & \tilde{E}_{n+1} \\ 0 & 0 & 0 \\ \tilde{E}^T_{n+1} & 0 & \tilde{K}_L \end{bmatrix} \begin{bmatrix} U_{n+1} \\ 0 \\ U_L \end{bmatrix} \tag{11}$$

5.    Increment n and return to step 1.  After all levels are reduced, $\tilde{K}_L$ is the lateral stiffness of the frame and $\bar{F}_L$ are the loads in the lateral direction acting on the floor systems.

After this process is completed for each frame substructure the frame stiffnesses and forces are transformed to the common global floor displacements and the three-dimensional equilibrium equations for the building are formed as:

$$K_g U_g = R_g + F_g \tag{12}$$

where $U_g$ contain  two translations and one rotation at each floor level, $R_g$ are the externally applied loads at the floor levels and $F_g$ are forces which are produced by loads on all frame substructures.  Equation (12) is a small dense system, three times the number of stories, which can be

solved directly. The lateral displacements and member
forces in each frame are then recovered by backsubstitution.

This approach is also very effective for the dynamic or
earthquake analysis of buildings since the mass of the
building can be lumped at the center of mass at each floor
level. Therefore only a relatively small eigenvalue need
be solved.

5. Solution of Equilibrium by Iteration. Iteration
methods were used extensively for the analysis of complex
structures prior to the availability of modern digital computers. Many early computer analysis programs were based on
iteration methods; however, for linear systems they are not
generally used for the following reasons:

1. Over-relaxation techniques based on physical insight,
which were very effective for hand calculation methods,
have been very difficult to generalize for arbitrary structural systems. In addition, the error correcting advantage
of iteration methods for hand calculations is of little
value if modern, highly reliable computers are used.

2. For most structural problems direct methods, which
bypass all zero operations, are faster and involve fewer
numerical operations. Also, multiple load conditions,
which are very common for practical problems, are solved
with very little extra numerical effort, whereas iteration
methods require a complete solution for each loading condition.

3. It is practically impossible to predict the number of
iterative cycles and computer time required to solve a
large structure, whereas the computer time required by a
direct method can be easily predicted within a few percent.

In order to illustrate the problems associated with an
iterative solution of a structural problem let us consider
the simple two-degree-of-freedom system shown in figure 8.
The equilibrium equations are given exactly to five significant figures as

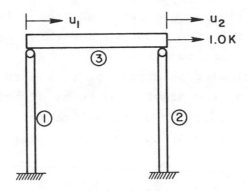

(a) TWO DOF FRAMES

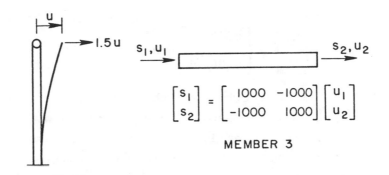

MEMBERS 1 & 2

MEMBER 3

(b) MEMBER STIFFNESS PROPERTIES

FIGURE 8   Structure to Illustrate Numerical
Sensitivity Problems

$$\begin{bmatrix} 1001.5 & -1000.0 \\ -1000.0 & 1001.5 \end{bmatrix} \begin{bmatrix} u_1 \\ u_2 \end{bmatrix} = \begin{bmatrix} 0.0 \\ 1.0 \end{bmatrix} \tag{13}$$

If eight significant figures are used in the solution, a
solution accurate to five significant figures is given as
$u_1$ = 0.33308 and $u_2$ = 0.33358.  If five significant figures
are retained in the stiffness matrix and solution procedure
the solution will be $u_1$ = 0.33283 and $u_2$ = 0.33333, or only

three significant figure accuracy in the solution. If four
significant figures are retained, the solution is
$u_1 = 0.4995$ and $u_2 = 0.5000$. For three significant figures
a solution is not possible.

The numerical sensitivity of this form can be eliminated
by formulating the problem in terms of absolute and relative
displacements. If the unknowns are selected as $u_1$ and $\Delta$
in which $u_2 = u_1 + \Delta$, the resulting equilibrium equations
are

$$\begin{bmatrix} 3.0 & 1.5 \\ 1.5 & 1000.5 \end{bmatrix} \begin{bmatrix} u_1 \\ \Delta \end{bmatrix} = \begin{bmatrix} 1.0 \\ 1.0 \end{bmatrix} \tag{14}$$

The solution to this equation, using five significant
figures, is

$$\begin{bmatrix} u_1 \\ \Delta \end{bmatrix} = \begin{bmatrix} 0.33308 \\ 0.00050 \end{bmatrix} \quad \text{and} \quad u_2 = 0.33358 \ .$$

Also, it is important to note that if two significant figures are retained and used, the solution is $u_1 = 0.33$ and
$\Delta = 0.00050$. This clearly illustrates the numerical advantages of the formulation which is a combination of absolute
and relative displacements.

The absolute and relative formulation has more significance for an iterative approach. For example the solution
of a two-degree-of-freedom system shown in figure 8 requires
1380 cycles with no over-relaxation. Figure 9 indicates the
number of cycles for convergence and illustrates the sensitivity of the method to the selection of the over-relaxation
factor. Of further interest is the fact that the solution
for the convergence limit of

$$\left| \frac{\delta u_1}{u_1} \right| + \left| \frac{\delta u_2}{u_2} \right| < .0001$$

produced results to only two significant figures. The 45
cycles required at the optimum over-relaxation factor of
1.92 illustrates the serious problem which can be encountered

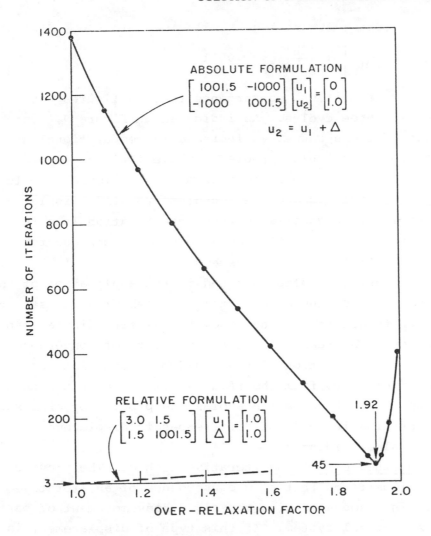

FIGURE 9   Effect of Over-Relaxation Factor
on Rate of Convergence

by the direct use of an iterative solution method.  One may
also draw the conclusion that equilibrium equations which
are numerically sensitive are also very slow to converge by
an iterative solution method.

Of major interest, however, is the application of iteration to the solution of Eq. (14) which is formulated in
terms of relative degrees of freedom.  For a convergence
limit of

$$\left|\frac{\delta u_1}{u_1}\right| + \left|\frac{\delta \Delta}{\Delta}\right| < .0001$$

and no over-relaxation, convergence to five figures was obtained in three cycles. As indicated in figure 9, over-relaxation factors caused an increase in computational effort for this particular problem. While it is not appropriate to draw general conclusions from such a simple example the apparent advantages of the absolute-relative displacement formulation suggest a further investigation into the appropriateness of iteration as a solution technique for more general finite element systems.

The impressive results obtained in the application of the iterative solution approach to this problem provide strong motivation to generalize the absolute-relative displacement formulation. The reason for the rapid rate of convergence is due to the separation of the rigid body translations of a group of elements from the relative displacements within the group. This formulation avoids the problem of expressing element strains in terms of differences in displacements which are almost identical.

6. Combined Direct and Iterative Method. The previous example illustrates that slow convergence of an iterative solution is associated with the rigid body movement of parts of the structural system. If this type of displacement is defined by a coarse mesh representation and the displacement of a fine mesh model is used to define relative displacement, then it is possible to generalize the absolute-relative formulation [8].

A three-dimensional finite element mesh is shown in figure 10. For this case the 16 nodes of the coarse mesh define the positions where the 48 absolute displacements are defined.

Within the coarse mesh element the absolute displacements may be expressed as:

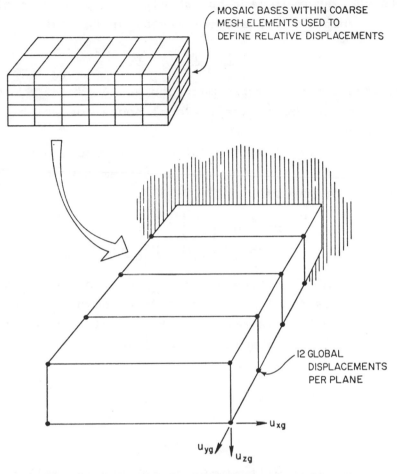

FIGURE 10    Use of Coarse Mesh to Define
Rigid Body Displacements

$$U_x = \sum_{j=1}^{8} H_j U_{jxg} + \Delta_x \quad , \quad U_y = \sum_{j=1}^{8} H_j U_{jyg} + \Delta_y \quad ,$$

$$U_z = \sum_{j=1}^{8} H_j U_{jzg} + \Delta_z \tag{15}$$

where $H_j$ are the natural interpolation functions for the
8-node isoparametric element.  If the fine mesh mosaic is
defined at convenient natural coordinate points $(r_i, s_i, t_i)$

within the large element the absolute displacement of any
node i within a large element may be written as:

$$\underline{U}_i = \underline{a}^* \, \bar{\underline{U}} = \underline{T}^* \, \underline{U}_g + \underline{\delta} \, \underline{\Delta}_i \tag{16}$$

where $\underline{a}^*$ and $\underline{T}^*$ are $3 \times 24$ matrices and $\delta$ is a $3 \times 3$ identity
matrix for all nodes except global nodes for which $\underline{\delta}$ is a
null matrix.

Based on this definition of node displacements the equilibrium equation for the complete system may be written in
submatrix form:

$$\begin{bmatrix} K_{rr} & K_{rg} \\ K_{gr} & K_{gg} \end{bmatrix} \begin{bmatrix} \Delta \\ U_g \end{bmatrix} = \begin{bmatrix} P_r \\ P_g \end{bmatrix} \tag{17}$$

The total number of free nodes for the fine mesh representation of this model is 336 or 1008 unknown displacements.
For this example $\Delta$ is a vector of 960 relative displacements
and $U_g$ contains 48 absolute displacements. Equation (17)
can be written as separate equations:

$$K_{rr} \, \Delta = P_r - K_{rg} \, U_g \tag{18}$$

and

$$K_{gg} \, U_g = P_g - K_{gr} \, \Delta \tag{19}$$

Since Eq. (19) will always be relatively small and contains the basic large translational and rotational displacements of the structure, it can be solved best with minimum
computational effort by a direct solution method. Equation
(18) contains a large number of degrees of freedom and will
be solved by iteration. In order to minimize computer time
and storage requirements the large matrix multiplications
suggested by Eqs. (18) and (19) should not be performed
directly, but should be performed at the element level.
A summary of a possible algorithm for the complete solution
is given in Table 2.

For many problems which have a large number of elements
with identical stiffnesses, it is apparent that storage
requirements for the algorithm given in Table 2 are minimal.

TABLE 2   Algorithm for Iterative Solution of
Finite Element Systems

1. Initial Calculations:

   a. Form element stiffnesses $K_m$ and transformation
      matrix relating element node displacements to
      global node displacements $T_m$.

   b. Evaluate diagonal terms of $K_{rr}$ and define as
      vector d.

   c. Form total stiffness with respect to global nodes

   $$K_{gg} = \sum_m T_m^T K_m T_m \quad .$$

   d. Form load vectors $P_r$ and $P_g$.
   e. Triangularize $K_{gg} = L D L^T$

2. Direct Solution of $K_{gg} u_g = P_g - K_{gr} \Delta$ :

   a. Form global loads $P_g^* = P_g - \sum_m T_m^T K_m \Delta^{(s)}$ .

   b. Forward reduce loads and back substitute to solve
      for approximate $u_g$ :  $L D L^T u_g^{(s)} = P_g^*$

3. Solve by Iteration for Relative Displacements
   $K_{rr} \Delta = P_r - K_{rg} u_g$ :

   a. Evaluate Solution Error
      $$E^{(s)} = P_r - K_{rg} u_g^{(s)} - K_{rr} \Delta^{(s)} \quad \text{or}$$
      $$E^{(s)} = P_r - \sum_m K_m [\Delta^{(s)} + T_m u_g^{(s)}] \quad \begin{array}{l}\text{summation over}\\ \text{elements}\end{array}$$

   b. Solve for Change in Relative Displacements
      $$\delta\Delta_i^{(s)} = E_i^{(s)}/d_i \ , \ i = 1 \text{ to total no. of displ.}$$

   c. Evaluate New Approximate Relative Displacement
      $$\Delta^{(s+1)} = \Delta^{(s)} + \beta\delta\Delta^{(s)} \ , \quad \text{where } \beta \text{ is a possible}$$
      over-relaxation factor.

   d. Check for convergence   $\sum_i |E_i^{(s)}|/\sum_i |R_i| < \varepsilon$ .
      Return to Step 2, Step 3, or stop iteration.

It may therefore be possible to solve rather large systems within core storage. In addition, the $T_m$ matrix for each element is extremely simple and can be formed from basic coordinate data when it is required rather than use high-speed or secondary storage. Even if secondary storage is required data can be transferred in large blocks of element stiffnesses. The number of cycles required in Step 2 or Step 3 must be established after some experience with the algorithm. The use of an over-relaxation factor must be investigated in both Step 2 and Step 3.

At the present time combined direct and iterative methods have not been used extensively in structural analysis. However, experience in other fields indicates the potential of the method [9]. Also, the algorithm may be appropriate to be used with new types of computer hardware which are based on arrays of microprocessors. In addition, the method may be effective in certain types of nonlinear problems and in programs that have adaptive mesh refinement.

7. Adaptive Mesh Refinement. Recent developments indicate that automated mesh refinement by a computer is now possible [10]. As illustrated in figure 11, the technique basically involves defining the geometry, loading and material properties of a structure with a coarse mesh. A preliminary coarse mesh analysis will indicate the areas of maximum stresses or elements with the maximum discontinuities. These critical elements are then subdivided and a new analysis is performed. This procedure can be repeated until the stress discontinuities have been reduced to a specified level. Since the technique involves a new solution of equations after each refinement this phase can dominate the numerical effort required unless special solution methods are used.

One approach is to use the direct-iterative approach presented in the previous section. For this case the additional relative displacements of the new elements would be selected as the correction to the previous mesh. This would

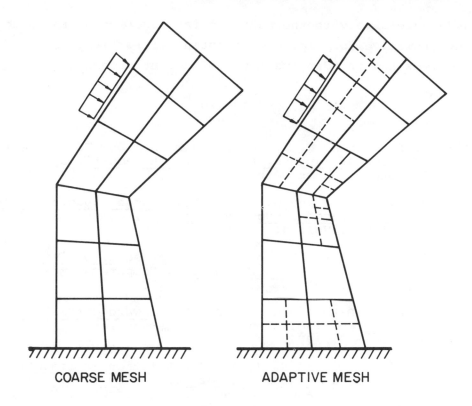

COARSE MESH          ADAPTIVE MESH

FIGURE 11   Example of Adaptive Mesh Refinement

utilize the solution from the previous mesh and only a few
relative displacements and the global displacements would
be altered during the new iterative solutions.

Another approach to the solution of equations in an
automated mesh refinement program is to form the new element
equations as corrections to the previous mesh.  If these
equations are added at the end of the previously factored
stiffness matrix it is possible to continue the factorization on the new equations only and then backsubstitute to
obtain all displacements.

8.  Summary.  At the present time there is not one method
for the solution of equations which is best for all structures.  The most significant saving is obtained if identical
parts of the structure are recognized in the basic formulation.  This can be accomplished in both the profile and
frontal solutions.  While iterative methods have not been

used extensively in the past the introduction of the absolute-relative formulation has potential for large bandwidth problems and in adaptive mesh refinement programs.

## REFERENCES

[1]   E. L. WILSON, K. J. BATHE and W. P. DOHERTY. Direct solution of large systems of linear equations, Computers and Structures, Vol. 4, pp. 363-372, 1974.

[2]   D. P. MONDKAR and G. H. POWELL. Large capacity equation solver for structural analysis, Computers and Structures, Vol. 4, pp. 699-728, 1974.

[3]   C. A. FELIPPA. Solution of linear equations with skyline-stored symmetric matrix, Computers and Structures, Vol. 5, pp. 13-29, 1975.

[4]   D. P. MONDKAR and G. H. POWELL. Toward optimal in-core equation solving, Computer and Structures, Vol. 4, pp. 531-548, 1974.

[5]   E. L. WILSON and H. H. DOVEY. Solution or reduction of equilibrium equations for large structural systems, Journal of Advances in Engineering Software, Vol. 1, 1978.

[6]   B. M. IRONS. A frontal solution program for finite element analysis, International Journal of Numerical Methods, Vol. 2, 1970.

[7]   E. L. WILSON and A. HABIBULLAH. TABS77 - A computer program for three-dimensional static and dynamic analysis of multistory buildings, Structural Mechanics Software Series, Vol. 2, University of Virginia Press, Charlottesville, 1978.

[8]   E. L. WILSON. Special numerical and computer techniques for the analysis of finite element systems, U.S.-German Symposium on Formulations and Computational Algorithms in Finite Element Analysis, M.I.T. Press, 1976.

[9]   E. L. WACHSPRESS. Two-level finite element computation, U.S.-German Symposium on Formulations and Computational Algorithms in Finite Element Analysis, M.I.T. Press, 1976.

[10]  A. PEANO et al. Adaptive approximations in finite element structural analysis, Computers and Structures, Vol. 10, pp. 333-342, 1978.

# Linear Programming Problems
# Arising from
# Partial Differential Equations

## Dianne P. O'Leary*

Abstract. In this paper we discuss the effectiveness of various linear programming algorithms on problems in which the model involves an underlying partial differential equation. Stable simplex implementations are the best general purpose algorithms, but a new variant of Gill and Murray's algorithm which uses iterative linear systems solvers rather than matrix updating methods is an alternative when storage is limited.

1. Introduction. The linear programming problem

$$
(1) \quad
\begin{array}{l}
\max b^T x \\
A_1 x = c_1, \\
A_2 x \geq c_2
\end{array}
\qquad
A = \begin{pmatrix} A_1 \\ A_2 \end{pmatrix}
\quad , \quad
c = \begin{pmatrix} c_1 \\ c_2 \end{pmatrix}
$$

where $A$ is an $m \times n$ matrix, $x$ is an $n \times 1$ vector, $c_1$ is an $m_1 \times 1$ vector, $c_2$ is an $m_2 \times 1$ vector, and $b$ is an $n \times 1$ vector has been thoroughly studied in many contexts. The simplex algorithm and some variants are effective for matrix problems in which $A$ is small and dense [15] . Using modifications such as the product [16] or elimination form of the inverse [25] large systems can also be solved. These methods were conceived for matrices arising from applications in scheduling, blending, and other areas in economics and business, and they work well. Structured matrices such as block or staircase can also be accommodated. Most of the popular algorithms are not numerically stable, but stable forms of the simplex algorithm have been proposed for dense in-core implementation [7] , sparse in-core [8,27] , and large out-of-core applications [29,30] .

However, these techniques are not always well adapted for problems in other application areas. For example, none of these algorithms take full advantage of the structure of the matrix $A$ when it arises from

*Department of Mathematics, University of Michigan, Ann Arbor, MI 48109.

discretization of a partial differential equation or finite element analysis of a structure. Typically, such problems are solved in a rather costly manner using existing general purpose linear programming algorithms. These matrices often have a block band structure with sparse blocks. Standard Gaussian elimination is often inefficient for linear systems arising from partial differential equations because of the large number of new nonzero elements introduced in the course of the algorithm; the standard simplex implementations have the same disadvantage. In addition, these matrices are often ill-conditioned, and safeguards should be incorporated to preserve the accuracy of the solution.

Much effort has been devoted to developing efficient and reliable algorithms which exploit the structure of linear systems of equations arising from partial differential equations. Such specialized methods include, for example, nested dissection [18] , relaxation techniques [31] , and the preconditioned conjugate gradient algorithm [3,12,22].

The purpose of this work is to develop linear programming algorithms which are adapted to problems related to elliptic partial differential equations and to determine their effectiveness. The tools used are ideas drawn from both mathematical programming and numerical analysis. The methods are simplex based but analogous to those used to find approximate solutions to the linear systems mentioned above, and thus the interface may be more natural for the user than that in typical linear programming codes. The algorithms are designed to be alternatives to standard linear programming packages and are more efficient in storage for this class of problems. Other features include the ability to handle inequality constraints without slack variables, limited modes of access to the original matrix A, and the ability to keep less than n constraints active at intermediate stages.

In Section 2 we review the meaning of an approximate solution to a linear programming problem. In Section 3 we discuss the application of various algorithms to linear programming problems and focus on a single algorithm derived from the simplex based method of Gill and Murray [20 ] . An applications-oriented reader might wish to begin with Section 4 where some examples of linear programming problems with constraints derived from partial differential equations are given.

In Section 5 we present some experimental results and conclusions.

2. Sensitivity of the Linear Programming Solution. In the problems we are considering, error is present in some of the data for the problem due to discretization of a partial differential equation. For example, in problem (1), if constraints arise from a second order accurate approximation to a partial differential equation, the components of $c$ may have errors proportional to the mesh size squared. Because of this, it is reasonable in computation to allow solutions $x$ which do not satisfy the constraints exactly. Thus, as long as our computed solution $x_c$ satisfies

$$A_1 x_c = c_1 + \Delta c_1$$
$$A_2 x_c \geq c_2 + \Delta c_2$$

for some $\Delta c_1$ and $\Delta c_2$ whose sizes are of order mesh size squared, $x_c$ can be considered feasible. Let $A_J$ be the matrix corresponding to those constraints which $x_c$ satisfies as equalities. Then, by the standard optimality condition for linear programming [15], if there is a vector $u$ which satisfies

$$A_J A_J^T u = -A_J (b + \Delta b)$$
$$u_j \geq 0 \qquad \text{if } j > m_1$$

where $\Delta b$ is of the same size as the uncertainty in $b$, then $x_c$ may be considered optimal. Assuming that the unperturbed problem has the same set of active constraints, the error incurred is $\Delta x = A_J^{-1} \delta c_J$ in $x$ and $b^T A_J^{-1} \delta c + \delta b^T x_c$ + higher order terms in the objective function value, where $\delta c_J$ ($\delta b$) is the difference between the true and the computational $c_J$ ($b$) vector.

3. Linear Programming Algorithms. There are two extremes in linear programming algorithms. The simplex method is an algorithm which, in exact arithmetic, is finitely terminating and gives the exact solution. Since an optimal solution to a linear programming problem can always be found at a vertex of the set of feasible points, the strategy is to step from vertex to adjacent vertex, increasing the objective function

value at each step. This is accomplished by storing and updating a basis matrix, a representation of the inverse of the matrix corresponding to the constraints satisfied as equalities at the current vertex. This algorithm is quite efficient in execution time, but the accessing and updating of the constraint bases can be costly if the use of secondary storage devices becomes necessary.

At the other extreme is a low storage but computationally intensive algorithm discussed, for example, in [13]. This involves applying the Agmon-Motzkin-Schoenberg [1,26] relaxation algorithm to the system of $m_1 + n + 2m_2 + 1$ equations and inequalities given by the primal-dual formulation for our linear programming problem. This algorithm requires only a few vectors of storage plus some representation of the original data matrices, but experiments on problems in the class we are considering here showed the computational requirements to be excessive even for very small problems [13] .

Thus, in this section, we discuss an algorithm which has a control structure similar to that of the simplex method but does not require as much storage.

3.1.  Gill and Murray Linear Programming Algorithm. In the following algorithm we use the notation $J$ and $S$ to denote subsets of the set of integers between $1$ and $m = m_1 + m_2$, $J'$ to denote the set of elements in the set $J$ which are greater than $m_1$, card $J$ to denote the number of elements in $J$, $J \setminus \{s\}$ to denote the set of all elements of $J$ except $s$, $\bar{J}$ to denote the integers between $1$ and $m$ not in $J$, and arg max (arg min) to denote a value in the domain of a given function which maximizes (minimizes) it. The vector whose components are all equal to $1$ will be denoted by $e$. The matrix $A_J$ denotes the submatrix of $A$ formed from rows indexed by elements of $J$. Subvectors of $c$ are defined similarly. The j-th row of $A$ will be denoted by $a_j$.

This algorithm has a very simple geometric interpretation. Given a feasible point, not necessarily a vertex of the convex feasible set, the algorithm generates uphill directions and step lengths until the optimality condition is satisfied. This condition is that the gradient of the objective function is a linear combination of the gradients of the equality constraints plus a nonpositive linear combination of the gradients of the active inequality constraints:

$$-A_J^T u = b$$

$$u_{J'} \geq 0$$

where $A_J$ is the submatrix of $A$ formed from the rows with zero residuals:

$$A_J x = c_J.$$

At each iteration, the optimality condition is checked, and, if any component of $u_{J'}$ is negative, a corresponding constraint is dropped from the set of active ones. A step is then generated which keeps all other active constraints and increases the value of the objective function. This process is repeated until convergence.

Started from a nonfeasible point, the algorithm proceeds in the same way, but with the objective function $\bar{b}^T x$ where $\bar{b}$ is the sum of gradients of the violated constraints. The vector $\bar{b}$ is updated as more constraints become satisfied.

We now state the algorithm in more detailed form. This basic scheme was presented by Gill and Murray in [20], but the version given here handles equality as well as inequality constraints.

-------------------------------------------------------------

Step (0): Choose the initial guess $x^{(0)}$, the convergence tolerance $\varepsilon$, and the degeneracy tolerance $\varepsilon_{min}$, and set $k = 0$.

Step (1): In preparation for the iteration, we calculate the residual $r$, and the index sets $J$ for the active constraints and $S$ for the violated constraints:

$$r^{(\hat{k})} = Ax^{(\hat{k})} - c$$

$$J = \{j: \left| r_j^{(\hat{k})} \right| \leq \varepsilon \}$$

$$S = \{s \leq m_1: \left| r_s^{(\hat{k})} \right| > \varepsilon \} \cup \{s > m_1: r_s^{(\hat{k})} < -\varepsilon \}$$

If card $J > n$, then if $\varepsilon \leq \varepsilon_{min}$ then halt with a degenerate vertex; if $\varepsilon > \varepsilon_{min}$ then refine $\varepsilon$ and restart this step. If $S = \phi$, then let $\bar{b} = b$; otherwise a "Phase 1" iteration is necessary with

$$\bar{b} = -\sum_{j \in J} a_j \; sgn(r^{(\hat{k})})$$

Step (2): For $k = \hat{k}, \hat{k}+1, \ldots$
(a) Calculate Lagrange multipliers $u_J$ satisfying

(*)
$$-A_J A_J^T u_J = A_J \bar{b}$$

Case 1: If $u_{J^i} \geq \epsilon \; e$ and card $J < n$, there is no constraint to drop. Set $s = 0$ and go to (b).

Case 2: If $u_{J^i} \geq \epsilon \; e$ and card $J = n$, the basis is full and there are no constraints to drop. Refine $\epsilon$. If $\epsilon \geq \epsilon_{min}$, then set $\hat{k} = k$ and return to Step (1). Else halt: if $S = \phi$, an optimal solution has been found. If $S \neq \phi$, no feasible solution exists.

Case 3: Otherwise, drop a constraint $s$ where $s = \arg \max_{j \in J'} -u_j$. Replace $J$ by $J \setminus \{s\}$.

(b) Calculate a feasible and uphill direction $p^{(k)}$ where

$$A_J p^{(k)} = 0$$

(**)
$$a_s^T p^{(k)} \geq 0$$

$$\bar{b}^T p^{(k)} > 0.$$

(c) Calculate the maximum steplength which does not add constraint indices to $S$. The first constraint encountered in direction $p^{(k)}$ is $q$ where

$$q = \arg \min_{j \notin J} (-r_j / a_j^T p^{(k)})$$

$$a_j^T p^{(k)} < 0$$

and the steplength is $\alpha_k = -r_q / a_q^T p^{(k)}$.
(d) Update the iterate , residual, and active index set:

$$x^{(k+1)} = x^{(k)} + \alpha_k p^{(k)}$$

$$r_{\bar{J}}^{(k+1)} = r_{\bar{J}}^{(k)} + \alpha_k A_{\bar{J}} p^{(k)}$$

$$r_J^{(k+1)} = r_J^{(k)}$$

$$J = J \cup \{q\}.$$

(e) If $q \notin S$, the violated set has not changed; go on to the next value of $k$. Otherwise, constraint $q$ is removed from the set $S$ and, if possible, the direction $p^{(k)}$ is reused for another step:

$$S = S \setminus \{q\}, \qquad \bar{b} = \bar{b} + a_q \; sgn(r_q^{(\hat{k})})$$

If $S = \phi$, Phase 1 is completed and we let $\bar{b} = b$. If $\bar{b}^{-T} p^{(k)} < 0$ or $q < m_1$, go on to the next value of $k$. Otherwise, let $J = J \setminus \{q\}$, $x^{(k)} = x^{(k+1)}$, $r^{(k)} = r^{(k+1)}$, and return to step (c).

-----------------------------------------------------------------

Standard modifications can be made to prevent cycling in the case of degeneracy.

Two of the complex computational subproblems in the algorithm are marked (*) and (**). Gill and Murray note that one solution to (**) is

$$p = (A_J^T (A_J A_J^T)^{-1} A_J - I) \bar{b}.$$

Thus we have two linear systems to solve at each iteration, both involving the matrix $A_J A_J^T$, although $I$ may change. The formulation of the problem as (1) rather than the more standard equality constraint problem allows us to keep less than a full basis on intermediate steps, and thus reduces the size of the linear systems. This is a decided advantage.

The standard approach to solving these linear systems is through the storage and updating of some representation of the factors or inverse of the matrix involved. For full matrices $A$ this is certainly the most efficient, but when $A$ is large and sparse, some alternatives, hinted at in [19], are available. A variety of methods can be used to solve the linear system, but one property is essential in a practical algorithm: it must not require $A_J A_J^T$ to be formed explicitly, since this matrix can be quite dense even when $A_J$ is sparse.

The method discussed by Gill and Murray is a decomposition of $A_J$ into the product of an orthogonal matrix $Q$ and a right triangular matrix $R$:

$$A_J^T = Q \begin{pmatrix} R \\ 0 \end{pmatrix}$$

Then $A_J A_J^T = R^T R$ and linear systems involving this matrix can be solved by forward and backward substitution. The triangular factors can be updated easily as the basis changes [ 21 ], although the sparsity structure of $A_J$ cannot be expected to be preserved.

### 3.2. Alternative Solution Algorithms for the Linear Systems.

If storage is at a premium then iterative algorithms for linear systems are an attractive alternative. The computed solution to (*) need not be very accurate, since only the signs of the components are essential. An accurate solution to the second system is more crucial since it is a projection of $\bar{b}$ on the active constraints, but inaccuracies here could be corrected by a later refining step, solving

$$A_J^T A_J x = A_J^T c_J, \qquad A_{\bar{J}} x \geq c_{\bar{J}} \, ,$$

as long as the final objective function value is greater than the one before the step.

In this section we consider, as an example, a particular iterative method, the preconditioned conjugate gradient algorithm [3,12,22 ]. This algorithm will compute a least squares solution to the problem $Fy \approx d$ (i.e., $F^T Fy = F^T d$) where $F$ and $d$ are given, and the pre-conditioning can be chosen in order to speed convergence.

The i-th step of the preconditioned conjugate gradient algorithm is an efficient organization of the computation involved in minimizing $\| Fy - d \|_2$ over all vectors of the form $y^{(0)}$ plus a vector in the span of $\{ C^{-T} C^{-1} F^T r^{(0)}, \ C^{-T} B C^{-1} F^T r^{(0)}, \ldots, \ C^{-T} B^{i-1} C^{-1} F^T r^{(0)} \}$ where $C$ is the chosen preconditioning operator, $B = C^{-1} F^T F C^{-T}$, and $r^{(0)} = d - F y^{(0)}$. The convergence rate is governed by the eigenvalues of B, and, in general, it is desirable that they be clustered.

An implementation particularly adapted to our problem is the SSOR (symmetric successive overrelaxation) preconditioned conjugate gradient algorithm described by Björck and Elfving [9]. The scaling operator C is of the form $(D + \omega L) D^{-1/2}$ where $F^T F = L + D + L^T$, L is a strictly lower triangular matrix, and D is a diagonal matrix. This algorithm requires access to the matrix F only column by column, and storage requirements are limited to those necessary to store several vectors of dimension equal to the number of rows of F and to generate F by columns. The work per iteration is equivalent to two multiplications of a vector by the matrix F plus several vector operations. The algorithm is as follows:

Choose $y^{(0)}$ in the range of $F^T$ and set $r^{(0)} = d - Fy^{(0)}$ and $p^{(0)} = C^{-1}F^T r^{(0)}$. For $i = 0, 1, \ldots$ Compute the new iterate and residual using the search direction $p^{(i)}$ and compute a new search direction for the next iteration:

$$y^{(i+1)} = y^{(i)} + \alpha_i C^{-T} p^{(i)}$$

$$r^{(i+1)} = r^{(i)} - \alpha_i F C^{-T} p^{(i)}$$

$$p^{(i+1)} = C^{-1} F^T r^{(i+1)} + \beta_i p^{(i)}$$

where

$$\alpha_i = r^{(i)T} F C^{-T} C^{-1} F^T r^{(i)} \;/\; p^{(i)T} C^{-1} F^T F C^{-T} p^{(i)}$$

$$\beta_i = r^{(i+1)T} F C^{-T} C^{-1} F^T r^{(i+1)} / r^{(i)T} F C^{-T} C^{-1} F^T r^{(i)}$$

The vectors $t = C^{-T} p$, $h = F C^{-T} p$, and $w = C^{-1} F^T r$ are computed from the formulas

$$h^{(n')} = 0$$

For $j = n', n'-1, \ldots, 2, 1$

$$t_j = d_j^{-1/2} p_j - \omega f_{.j}^T h^{(j)}$$

$$h^{(j-1)} = h^{(j)} + f_{.j}t_j$$

$$h = h^{(0)}$$

$$g^{(0)} = r$$

For $j = 1,2,\ldots,n'$

$$w_j = d_j^{-1/2} f_{.j}^T g^{(j)}$$

$$g^{(j+1)} = g^{(j)} - (\omega d_j^{-1/2} w_j) f_{.j}$$

where $n'$ is the number of columns of $F$, $f_{.j}$ is its $j$-th column, and $d_j$ is the $j$-th main diagonal element of $F^T F$.

--------------------------------------------------------------------

Note that in our application, $f_{.j}$ is a row of the constraint matrix $A$, so access to the original matrix is always by row.

4. Sources of Problems with Differential Operator Constraints. Linear programming problems with constraints derived from partial differential equations arise in many contexts. Several examples are sketched below.

4.1. Optimal Pumping for Drainage [2]. One variant of this problem is to determine a pumping strategy to maintain the water level in a potential excavation site at or below a given depth $h_r$. If we denote by $v(x,y)$ the square of the hydraulic head at the point $(x,y)$, and by $w(x,y)$ the source or sink term for water flow, we obtain the constraints

$$v_{xx} + v_{yy} = 2w/k$$

$$0 \le v \le h_r^2$$

$$w \ge 0$$

where $k$ denotes the conductivity, and appropriate boundary conditions are also given. Under specified conditions, it may be desired, for example, to minimize the pumping, which is an integral of the $w$ term. After discretization by finite difference or finite element techniques, this problem becomes a linear programming problem.

4.2. Pollution Control [17]. In a typical problem of this type, the diffusion-convection equation is coupled with appropriate boundary and nonnegativity constraints and an objective of determining the maximum pollution outputs (either of a given effluent or thermal waste) which maintain a given level of water quality. After discretization, the resulting linear programming problem is of a form similar to the one described above.

4.3. Free Boundary Problems. Many free boundary problems such as determining the interface between the liquid and gas phases of lubricant in a journal bearing [14] or describing water flow through a porous dam [4] have been written as linear complementarity problems:

$$Mz + q = r$$
$$r^T z = 0$$
$$r \geq 0, z \geq 0,$$

where M arises from discretization of a partial differential equation. Mangasarian [23] has established that if M is a Z matrix, which is often the case in finite difference discretizations, then this problem can either be solved in its original form or cast as a linear programming problem with constraints

$$Mz + q \geq 0$$

and z nonnegative.

4.4 Solution of Differential Equations by Collocation [ 24,28 ]. If we choose to represent the approximate solution to a differential equation as a linear combination of n basis functions, and demand that this approximation exactly satisfy the differential equation or boundary conditions at n points, we obtain a square system of equations $Ax = b$. If we write the equations for more than n points, we obtain an over-determined system. Possible ways to solve this include solving $Ax + y = b$ subject to minimizing the $l_1$ norm or the $l_\infty$ norm of y. Either of these problems can be formulated as a linear programming problem. If the basis chosen is local (for example, a finite element basis), then under appropriate ordering of equations and unknowns, the matrix  A  will be sparse and structured.  Such systems might be solved more efficiently using an algorithm for solving overdetermined systems (see, for example, [ 5,6,10,11] ) rather than the simplex algorithm, but the approach developed here could be adapted for use in these algorithms, too.

5. Results and Conclusions.  The algorithm of Section 3 was implemented and tested on a series of problems related to the pumping model of Section 4.1.  The drainage site was taken to be a rectangle, and boundary conditions were given on the edges of a larger rectangle containing it.  A finite difference grid was imposed, and the discretization of the partial differential equation was performed using the second order accurate five point difference operator.  Let  $v^T = (v_1^T, v_2^T)$  where  $v_1$  corresponds to mesh points in the drainage site and  $v_2$  to points in the outer region.  A pumping vector  $w$  is partitioned similarly and no pumping is allowed in the drainage region.  Then the problem becomes

$$\max \quad -w_2^T e, \qquad e = (1,1,\ldots 1)^T$$

$$Av - 2/k \begin{pmatrix} 0 & 0 \\ 0 & I \end{pmatrix} w = c_1$$

$$0 \le v_1 \le h_r^2$$

$$v_2 \ge 0, \quad w_2 \ge 0$$

$$w_1 = 0$$

where  $c_1$  is a vector containing zeroes and the boundary conditions. Both the region and the boundary conditions were taken to be symmetric, so the problem was solved on a quarter domain.  This setup is similar to that discussed in [2 ].  A  is a sparse matrix with at most five nonzero elements per row.

There is no difference in the number of iterations needed for Gill and Murray's algorithm whether matrix updating techniques or iterative methods are used. On a small problem, for example ( 34 unknowns, 59 constraints), it took 15 iterations to reach the feasible region and 10 more to find the optimal point, using .1e as an initial guess. Therefore, the emphasis in experiments was on determining the effectiveness of the iterative algorithms on least squares problems with matrix columns corresponding to a subset of the linear programming constraint rows.

When conjugate gradients preconditioned with SSOR are used to solve linear systems involving matrices related to $\tilde{A}$, typically only a small number of iterations are required to obtain a good approximation to the solution vector. Unfortunately, this does not seem to be the case for least squares problems. Solving them by this algorithm seems to be a $O(n'^2)$ process. Theoretically the algorithm must terminate in at most $n'$ iterations with the exact solution, but it was observed that between $.5n'$ and $n'$ iterations were necessary to obtain a good approximation, and sometimes the algorithm failed to converge even in $n' + 10$ iterations due to round off error. Varying $\omega$ over the range .5 to 1.9 changed the number of iterations at most by a factor of two. Without scaling, the conjugate gradient algorithm often did not converge, indicating that these problems are indeed ill-conditioned. Double precision is necessary even for small problems.

In storage, of course, the iterative methods are vastly superior. This implementation required 16 vectors for both the linear programming iteration and the least squares problems, plus space required to generate A by row, and further savings could be achieved.

Although currently the iterative methods are not competitive in computation time with matrix updating techniques, their low overhead may make them an attractive alternative in some instances. In addition, the modular design of the algorithm permits the easy substitution of new least squares methods as they become available.

Acknowledgement.

This work was supported by National Science Foundation Grant MCS76-06595 at the University of Michigan.

# REFERENCES

1.  SHMUEL AGMON, The relaxation method for linear inequalities, Canadian J. on Math. 6 (1954), pp. 382-392.

2.  EDUARDO AGUADO, IRWIN REMSON, MARY F. PIKUL and WILL A. THOMAS, Optimal pumping for aquifer dewatering, J. of the Hydraulics Div. ASCE 100 (1974), pp. 869-877.

3.  OWE AXELSSON, Solution of linear systems of equations: iterative methods, in Sparse Matrix Techniques Copenhagen 1976, V.A. Barker, ed., Springer-Verlag, New York, 1977, pp. 1-51.

4.  C. BAIOCCHI, V. COMINCIOLI, E. MAGENES, and G. A. POZZI, Free boundary problems in the theory of fluid flow through porous media, Ann. Mat. Pura. Appl. 97 (1973), pp. 1-82.

5.  R. H. BARTELS, A. R. CONN and C. CHARLAMBOUS, On Cline's direct method for solving overdetermined linear systems in the $l_\infty$ sense, SIAM J. Numer. Anal. 15 (1978), pp. 255-270.

6.  R. H. BARTELS, A. R. CONN and J. W. SINCLAIR, Minimization techniques for piecewise differentiable functions: the $l_1$ solution to an overdetermined linear system, SIAM J. Numer. Anal. 15 (1978), pp. 224-241.

7.  R. H. BARTELS and G. H. GOLUB, The simplex method of linear programming using LU decomposition, Comm. ACM 12 (1969), pp. 266-268.

8.  R. H. BARTELS, G. H. GOLUB, and M. A. SAUNDERS, Numerical techniques in mathematical programming, in Nonlinear Programming, J. Abadie, ed., Academic Press, New York, 1970, pp. 123-176.

9.  ÅKE BJÖRCK and TOMMY ELFVING, Accelerated projection methods for computing pseudoinverse solutions of systems of linear equations, Linkoping Univ. Mathematics Dept. Rept. LiTH-MATH-R-1978-5, Linkoping, Sweden, 1978.

10. E. W. CHENEY, Introduction to Approximation Theory, McGraw-Hill Book Co., New York, 1966.

11. A. K. CLINE, A descent method for the uniform solution to overdetermined systems of linear equations, SIAM J. Numer. Anal. 13 (1976), pp. 293-309.

12. PAUL CONCUS, GENE H. GOLUB, and DIANNE P. O'LEARY, A generalized conjugate gradient method for the numerical solution of elliptic partial differential equations, in Sparse Matrix Computations, James R. Bunch and Donald J. Rose, eds., Academic Press, New York, 1976, pp. 309-322.

13. RICHARD W. COTTLE and JONG-SHI PANG, On solving linear complementarity problems as linear programs, Math. Programming Study 7 (1978), pp. 88-107.

14. C. W. CRYER, The method of Christopherson for solving free boundary problems for infinite journal bearings by means of finite differences, Math. Comp. 25 (1971), pp. 435-443.

15. G. B. DANTZIG, Linear Programming and Extensions, Princeton Univ. Press, Princeton, New Jersey, 1963.

16. G. B. DANTZIG and W. ORCHARD-HAYS, The product form for the inverse in the simplex method, Math. Comp. 8 (1954), pp. 64-67.

17. TANEHIRO FUTAGAMI, The finite element and linear programming method and the related methods, Ph.D. thesis, Hiroshima Inst. of Tech. Dept. of Civil Engin., April, 1976.

18. ALAN GEORGE, Nested dissection of a regular finite element mesh, SIAM J. Numer. Anal. 10 (1973), pp. 345-363.

19. P. E. GILL and W. MURRAY, Methods for large scale linearly constrained problems, in Numerical Methods for Constrained Optimization, P. E. Gill and W. Murray, eds., Academic Press, New York, 1974, pp. 93-148.

20. P. E. GILL and W. MURRAY, A numerically stable form of the simplex algorithm, Linear Alg. and Appl. 7 (1973), pp. 99-138.

21. P. E. GILL, G. H. GOLUB, W. MURRAY and M. A. SAUNDERS, Methods for modifying matrix factorizations, Math. Comp. 28 (1974), pp. 505-535.

22. M. R. HESTENES, The conjugate gradient method for solving linear systems, Proc. Symp. Appl. Math. VII Numer. Anal. (1956), pp. 83-102.

23. O. L. MANGASARIAN, Linear complementarity problems solvable by a single linear program, Math. Prog. 10 (1976), pp. 263-270.

24. O. L. MANGASARIAN, Numerical solution of the first biharmonic-problem by linear programming, Int. J. Engin. Sci. 1 (1963), pp. 231-240.

25. H. M. MARKOWITZ, The elimination form of the inverse and its application to linear programming, Man. Sci. 3 (1957), pp. 255-269.

26. TH. MOTZKIN and I. J. SCHOENBERG, The relaxation method for linear inequalities, Canadian J. on Math. 6 (1954), pp. 393-404.

27. J. K. REID, A sparsity-exploiting variant of the Bartels-Golub decomposition for linear programming bases, AERE Rept., Harwell, Didcot, Oxford, 1975.

28. J. B. ROSEN, Approximate computational solution of non-linear parabolic partial differential equations by linear programming, in Numerical Solutions of Nonlinear Differential Equations, Donald Greenspan, ed., John Wiley&Sons, New York, 1966, pp. 265-296.

29. MICHAEL A. SAUNDERS, <u>A fast, stable implementation of the simplex method using Bartels-Golub updating</u>, in Sparse Matrix Computations, James R. Bunch and Donald J. Rose, eds., Academic Press, New York, 1976, pp. 213-226.

30. M. A. SAUNDERS, <u>Product form of the Cholesky factorization for large scale linear programming</u>, Stanford Univ. Computer Science Dept. Rept. STAN-CS-72-301, 1972.

31. R. S. VARGA, <u>Matrix Iterative Analysis</u>, Prentice-Hall, Englewood Cliffs, New Jersey, 1962.

# Shifted Incomplete Cholesky Factorization

## Thomas A. Manteuffel*

Abstract.  This talk will describe a matrix splitting
known as the shifted incomplete Cholesky factorization.
After a brief discussion of the generalized conjugate
gradient method for accelerating the splitting, incomplete
Cholesky factorization will be described.  Next, sufficient
conditions for stable incomplete factorization will be
discussed.  The shifted incomplete Cholesky factorization
will then be introduced followed by a discussion of
several open questions involving this factorization.
Finally, a test problem will be used to reveal some of the
characteristic behavior of this method.

1.  Introduction.  This talk will demonstrate a method
for solving large, sparse, positive definite linear
systems.  This talk will be tutorial in nature.  I hope
to describe the shifted incomplete Cholesky factorization
and factorization procedures in general, to describe
some of the mathematics involved, and most importantly, to
expose some interesting and important open questions.

*Sandia Laboratories, Livermore, CA  94550.

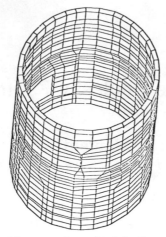

Figure 1. Cylinder

This work was motivated by a three-dimensional model of the structural deformation in a cylinder (see Figure 1) due to an applied force. The model produced a linear system with 18 000 unknowns, a half-bandwidth of 1300, and an average of 115 nonzeroes per equation. The direct solution of such a system would require approximately 15 000 000 000 arithmetic operations and 23 000 000 words of storage. Using the shifted incomplete Cholesky factorization followed by a generalized conjugate gradient iteration the system was solved in less than half the time required by the direct method and only 3 000 000 words of storage were required (see Manteuffel, [6]).

The shifted incomplete Cholesky factorization is a form of preconditioning or matrix splitting. Given the linear system

$$A\underline{x} = \underline{b},$$

A is split into two parts

$$A = M - N$$

where M is easily invertible. By this we mean that M is nonsingular and it is easy to solve systems of the type

$$M\underline{x} = \underline{y} .$$

We can _implicitly_ form the preconditioned system

$$M^{-1}A\underline{x} = M^{-1}\underline{b} .$$

If M is a "good" approximation of A, then $M^{-1}A$ will approximate the identity matrix and the preconditioned system will be easier to solve by iterative techniques than the original system.

If both M and A are positive definite, the preconditioned system may be solved by a variant of the conjugate gradient algorithm (Concus, Golub, and O'Leary,[1]). Otherwise, methods based upon the Tchebychev polynomials in the complex plane may be employed (Manteuffel, [5]). These methods are polynomial methods in that at each step the approximate solution is improved by adding a vector that is a linear combination of members of a Krylov sequence in $M^{-1}A$ (Rutishauser, [2]). Given an initial guess $\underline{x}_0$ let the residual be

$$\underline{r}_0 = \underline{b} - A\underline{x}_0 .$$

Let

$$\underline{h}_0 = M^{-1}\underline{r}_0$$

be the generalized residual and let

$$V_i = \{ \underline{h}_0, M^{-1}A\underline{h}_0, (M^{-1}A)^2\underline{h}_0, \ldots, (M^{-1}A)^i\underline{h}_0 \}$$

be the subspace spanned by the first i+1 terms of the

Krylov sequence of $\underline{h}_o$ and $M^{-1}A$. After i steps of iteration, a polynomial method will yield

$$\underline{x}_i = \underline{x}_o + \underline{p}_{i-1}$$

for some $\underline{p}_{i-1} \varepsilon V_{i-1}$. If $\underline{x}$ is the exact solution and

$$\underline{e}_i = \underline{x} - \underline{x}_i$$

is the error at step i, then

$$\underline{e}_i = \underline{e}_o - \underline{p}_{i-1} .$$

Since $\underline{p}_{i-1} \varepsilon V_{i-1}$, there is some polynomial $P_{i-1}(z)$ of degree i-1 such that

$$\underline{p}_{i-1} = P_{i-1}(M^{-1}A)\underline{h}_o = P_{i-1}(M^{-1}A)M^{-1}A\underline{e}_o .$$

Thus, the error at step i may be expressed as

$$\underline{e}_i = (I - P_{i-1}(M^{-1}A)M^{-1}A)\underline{e}_o .$$

The polynomial $P_{i-1}(M^{-1}A)$ is an approximation to $(M^{-1}A)^{-1}$. The rate at which the iteration converges depends upon how well a polynomial of degree i-1 can approximate $(M^{-1}A)^{-1}$. This depends mainly upon the condition of $M^{-1}A$; that is, the ratio of the largest eigenvalue of $M^{-1}A$ to the smallest eigenvalue of $M^{-1}A$:

$$C(M^{-1}A) = \frac{\lambda \max}{\lambda \min} .$$

More precisely, the rate of which the conjugate gradient iteration will converge depends upon all of the ratios

$$\frac{\lambda_i}{\lambda_j} , \quad i \geq j .$$

The best matrix splitting is the splitting that yields the best spectrum of $M^{-1}A$ (Greenbaum, [4]).

2. What is incomplete Cholesky factorization? In order to better understand incomplete factorization let us first examine complete Cholesky factorization. Given a positive

definite matrix $A = (a_{ij})$, write

$$A = A_1 = \begin{pmatrix} a_{11}^{(1)} & \underline{a}_1^T \\ \hline \underline{a}_1 & B_1 \end{pmatrix}$$

where $\underline{a}_1$ is the first column of A without the first element. Since A is positive definite, then the pivot element $a_{11} > o$ and so we may write

$$A_1 = \begin{pmatrix} 1 & \underline{o}^T \\ \hline \underline{\ell}_1 & I \end{pmatrix} \begin{pmatrix} a_{11} & \underline{o}^T \\ \hline \underline{o} & A_2 \end{pmatrix} \begin{pmatrix} 1 & \underline{\ell}_1^T \\ \hline \underline{o} & I \end{pmatrix}$$

$$= L_1 \, \Omega_1 \, L_1^T \, ,$$

where

$$A_2 = B_1 - \frac{1}{a_{11}} \underline{a}_1 \underline{a}_1^T \quad , \quad \underline{\ell}_1 = \frac{1}{a_{11}} \underline{a}_1 \quad .$$

Write

$$A_2 = \begin{pmatrix} a_{22}^{(2)} & \underline{a}_2^T \\ \hline \underline{a}_2 & B_2 \end{pmatrix} \quad .$$

If A is positive definite then $A_2$ will also be positive definite and in particular $a_{22}^{(2)} > o$. The submatrix $A_2$ can be factored in the same manner. We have

$$A_1 = \begin{pmatrix} 1 & & \underline{o}^T \\ \hline \underline{\ell}_1 & \begin{matrix} 1 & \underline{o}^T \\ \underline{\ell}_2 & I \end{matrix} \end{pmatrix} \begin{pmatrix} a_{11}^{(1)} & & \\ & a_{22}^{(2)} & \underline{o} \\ \hline & \underline{o} & A^3 \end{pmatrix} \begin{pmatrix} 1 & & \underline{\ell}_1^T \\ \hline \underline{o} & \begin{matrix} 1 & \underline{\ell}_2^T \\ \underline{o} & I \end{matrix} \end{pmatrix}$$

$$= L_2 \, \Omega_2 \, L_2^T \, ,$$

where

$$A_3 = B_2 - \frac{1}{a_{22}^{(2)}} \underline{a}_2 \underline{a}_2^T \quad , \quad \underline{\ell}_2 = \frac{1}{a_{22}^{(2)}} \underline{a}_2 .$$

Again the submatrix $A_3$ will be positive definite. Continuing in this manner we will eventually have

$$A_1 = L \, \Omega \, L^T, \quad \Omega = \mathrm{diag}(\rho_i), \quad \rho_i = a_{ii}^{(i)} > o .$$

Suppose A is sparse. At each step of the algorithm some of the sparsity is destroyed. For instance, we have

$$A_2 = B_1 - \frac{1}{a_{11}^{(1)}} \underline{a}_1 \underline{a}_1^T .$$

The rank one matrix $\underline{a}_1 \underline{a}_1^T$ may fill in some of the zeroes in $B_1$. The process of fill-in is cumulative and the final lower triangular matrix L may be much more dense than the lower triangular part of the matrix A.

With incomplete factorization the zero structure of the lower triangular L is determined beforehand and undesirable fill-in is discarded. We first split $A_1$,

$$A_1 = \left( \begin{array}{c|c} a_{11}^{(1)} & \underline{a}_1^T \\ \hline \underline{a}_1 & B_1 \end{array} \right) = \left( \begin{array}{c|c} a_{11}^{(1)} & \underline{b}_1^T \\ \hline \underline{a}_1 & B_1 \end{array} \right) - \left( \begin{array}{c|c} o & \underline{r}_1^T \\ \hline \underline{r}_1 & o \end{array} \right)$$

$$= M_1 - R_1 ,$$

where $\underline{b}_1$ is found by setting certain elements of $\underline{a}_1$ to zero. We then perform one step of complete factorization on $M_1$ to get

$$A_1 = \left( \begin{array}{c|c} 1 & \underline{o}^T \\ \hline \underline{\ell}_1 & I \end{array} \right) \left( \begin{array}{c|c} a_{11}^{(1)} & \underline{o}^T \\ \hline \underline{o} & A_2 \end{array} \right) \left( \begin{array}{c|c} 1 & \underline{\ell}_1^T \\ \hline \underline{o} & I \end{array} \right) - \left( \begin{array}{c|c} o & \underline{r}_1^T \\ \hline \underline{r}_1 & o \end{array} \right)$$

where

$$A_2 = B_1 - \frac{1}{a_{11}^{(1)}} \underline{b}_1 \underline{b}_1^T \quad , \quad \underline{\ell}_1 = \frac{1}{a_{11}^{(1)}} \underline{b}_1 \; .$$

Notice that the zero structure of $\underline{\ell}_1$ is dictated by the zero structure of $\underline{b}_1$. Also notice that $\underline{r}_1$ has zeroes where $\underline{\ell}_1$ has nonzeroes.

We next split $A_2$ by setting certain elements in the first row and first column of $A_2$ to zero. If the pivot element $a_{22}^{(2)} \neq o$, then we may again factor to get

$$A_1 = \left( \begin{array}{c|c|c} 1 & \underline{o}^T & \\ \hline \underline{\ell}_1 & 1 & \underline{o}^T \\ & \underline{\ell}_2 & I \end{array} \right) \left( \begin{array}{c|c} a_{11}^{(1)} & \\ & a_{22}^{(2)} & o \\ \hline & o & A_3 \end{array} \right) \left( \begin{array}{c|c|c} 1 & \underline{\ell}_1^T & \\ \hline \underline{o} & 1 & \underline{\ell}_2^T \\ & \underline{o} & I \end{array} \right) - \left( \begin{array}{c|c|c} o & \underline{r}_1^T & \\ \hline \underline{r}_1 & o & \underline{r}_2^T \\ & \underline{r}_2 & o \end{array} \right).$$

The zero structure of $\underline{\ell}_1$ and $\underline{\ell}_2$ have been determined by $\underline{b}_1$ and $\underline{b}_2$. Again notice that $\underline{r}_2$ has zeroes where $\underline{\ell}_2$ is nonzero.

We may continue in this fashion as long as the pivot elements $a_{ii}^{(i)} \neq o$. We have the splitting

$$A = L \sum L^T - R = M - R, \quad \sum = \text{diag} (\sigma_i) \quad , \quad \sigma_i = a_{ii}^{(i)} \; .$$

The lower triangular matrix L will have the zero structure we have dictated and the matrix R will contain the discarded fill-in. In practice, fill-in is never accumulated. Only those elements which are to be nonzero in the final L

are allowed to fill-in in the intermediate steps.  The
matrix R is not computed.

To better understand this splitting let us examine the
undirected graph of the matrix A; that is, a set of points
$p_i$, i = 1,...,n where n is the dimension of A together
with a set edge $\{p_i, p_j\}$ connecting $p_i$ to $p_j$ if and
only if $a_{ij} \neq 0$ or $a_{ji} \neq 0$.  Likewise, we may consider
the graph of the lower triangular matrix L.  Suppose we
choose the zero structure of L so that the graph of L
includes the graph of A as a subset; that is, if $a_{ij} \neq 0$
for some i > j, then $\ell_{ij} \neq 0$.  This implies that the
graph of A and the graph of R have no common edges.  We
know from above that $r_{ij} \neq 0$ for some i > j implies
$\ell_{ij} = 0$ which in turn implies $a_{ij} = 0$.  From the
expression

$$A = M - R$$

we see that the graph of M also contains the graph of A,
and in fact, if $a_{ij} \neq 0$ then $a_{ij} = m_{ij}$.  Thus, M
approximates A in that the graph of M looks like the graph
of A plus some extra edges.  Further, whenever A and M
have a common edge then the "magnitude" of the edges are
equal.  Figure 2 gives an example of the graphs of the
matrices A, L, R, and M.

It is most convenient to choose the graph of L to be
equal to the graph of A.  This allows the factorization to
be stored in the same data structure as A.  Let us call
this graph $G_0$.  It is sometimes desirable to allow more

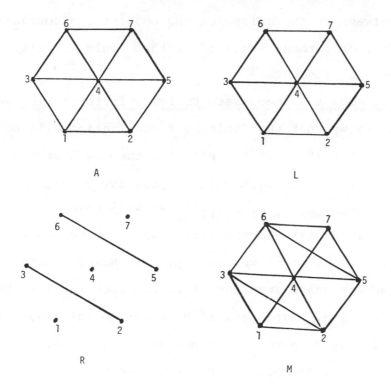

Figure 2.   Graphs of A, L, R, M

fill-in.   Consider the graph $G_1$ in which two points
share a common edge whenever there is a path of less than
two edges connecting the points in the graph $G_0$.   Let us
call this the extension graph of the first level.   General-
izing, one may consider the graph $G_i$ in which two points
share a common edge whenever there is a path of less than
$i + 1$ edges connecting the two points in $G_0$.   This
concept of extension reflects the heuristic that an
unknown is most strongly affected by its neighbors in the
graph.   Clearly, one may choose the graph to achieve any

balance between no factorization and complete factorization. In practice, only those edges of $G_i$ that would actually fill-in need be stored.

3. When does the incomplete Cholesky factorization work? It is well known that the Cholesky factorization will not breakdown. If A is positive definite, then each of the submatrices $A_i$, i = 1,...,n will be positive definite and thus for each pivot $\rho_i = a_{ii}{}^{(i)}$ we will have $\rho_i > o$. The incomplete factorization will not breakdown unless for some pivot $\sigma_i$ we have $\sigma_i = o$. However, the generalized conjugate gradient iteration requires M to be positive definite. Moreover, if M is not positive definite, then it will not be a very good approximation to A. Therefore we must have $\sigma_i > o$ for the pivots of the incomplete factorization as well. Having chosen a graph for L, we will say that the incomplete factorization, is stable if

$$\sigma_i > o \ , \quad i = 1,...,n \ .$$

Further, we cannot allow any pivot to be too small. A small pivot would make M almost singular and thus $M^{-1}A$ will have poor condition. The stability of the factorization can be measured if we notice that the splitting is invariant to diagonal scaling. If Q is any diagonal matrix and we let

$$\hat{A} = QAQ,$$

and then perform incomplete Cholesky decomposition to get

$$A = L \textstyle\sum L^T - R = M - R,$$

$$\hat{A} = \hat{L} \sum \hat{L}^T - \hat{R} = \hat{M} - \hat{R},$$

where L and $\hat{L}$ have the same graph, then

$$\hat{L} = QLQ^{-1},$$

$$\hat{\sum} = Q \sum Q,$$

$$\hat{R} = QRQ.$$

Thus,

$$\hat{M} = QMQ$$

and so

$$\hat{M}^{-1}\hat{A} = (Q^{-1}M^{-1}Q^{-1})(QAQ) = Q^{-1}M^{-1}AQ \cong M^{-1}A.$$

Since the generalized conjugate gradient iteration is sensitive to the spectrum of $M^{-1}A$, the procedure is invariant to diagonal scaling.

Consider the scaling

$$\hat{A} = D^{-1/2}AD^{-1/2}, \quad D = \text{diag } (..a_{ii}..).$$

Then, we have

$$\hat{A} = I - B,$$

where

$$b_{ii} = 0,$$

$$|b_{ij}| < 1, \quad i \neq j.$$

Notice that B is similar to the Jacobi splitting:

$$B \cong I - D^{-1}A.$$

If we perform incomplete factorization on A, then the pivots obey the relations

$$\sigma_1 = 1,$$

$$\sigma_i \leq 1, \quad i = 2...,n.$$

Thus,

$$S = 1/\min_i \sigma_i$$

yields a measure of the stability of the factorization.

There are two known sufficient conditions for stability of the incomplete factorization ([3], [6], [7], [8]). One is diagonal dominance.

Theorem:  If A is diagonally dominant and positive definite, then incomplete Cholesky decomposition with any graph of L is stable.

Another sufficient condition is a Stieltjes matrix.

Definition:  A matrix $A = (a_{ij})$ is a Stieltjes matrix if A is positive definite, $a_{ij} < o$ for $i \neq j$, A is nonsingular, and $A^{-1} \geq o$.

Theorem:  If A is a Stieltjes matrix, then incomplete Cholesky factorization with any graph of L is stable. Further, it will be more stable than complete factorization.

In fact, we have the following interesting result (Manteuffel, [6]).

Theorem:  If A is a Stieltjes matrix, then adding edges to the graph of L will improve the condition of $M^{-1}A$ (up to a factor of 2).

Necessary and sufficient conditions for stability are not known.  In the incomplete factorization certain off-diagonal elements of the submatrices are set to zero. This may cause the submatrix to be no longer positive definite.  If a submatrix is not positive definite, then there is no way to guarantee that subsequent submatrices will be positive definite.  In general, stability depends upon the graph of L.  If we allow enough fill-in, then the incomplete factorization will become complete factorization, which is stable for any positive definite matrix.

4.  The Shifted Incomplete Cholesky Factorization.  The linear systems that arise from structural mechanics are positive definite, but are not Stieltjes nor are they diagonally dominant.  It was found through experiment that the incomplete factorization with the graph of L equal to the graph of A was unstable for many systems.  To overcome this problem one might increase the graph of L and allow more fill-in.  This proved to be inefficient.  As an alternative, consider the incomplete factorization of a matrix close to A but more nearly diagonally dominant. Suppose we have the diagonally scaled matrix

$$A = I - B .$$

Consider the pencil of matrices

$$A(\alpha) = I - \frac{1}{1+\alpha} B .$$

Clearly, for $\alpha$ sufficiently large $A(\alpha)$ is diagonally dominant.  Thus, there is some $\alpha \geq o$ for which the incomplete factorization of $A(\alpha)$ with a given graph of L is stable.  Suppose we have

$$A(\alpha) = L \sum L^T - R(\alpha) = M(\alpha) - R(\alpha) .$$

(Here L and $\sum$ also depend upon $\alpha$).  Then, we may write

$$I - \frac{1}{1+\alpha} B = M(\alpha) - R(\alpha)$$

and

$$I - B = M(\alpha) - (R(\alpha) + \frac{\alpha}{1+\alpha} B)$$

or

$$A = M(\alpha) - N(\alpha) .$$

For each $\alpha$ for which the incomplete factorization of $A(\alpha)$ is stable we have a splitting of A.

Although we are interested in the factorization for small $\alpha$, examining the asymptotic effect for large $\alpha$ is instructive. Notice that as $\alpha$ gets large $A(\alpha)$ approaches I. One would expect that $M(\alpha)$ also approaches I. In fact, if the graph of L includes the graph of A we have

$$M(\alpha) = I - \frac{1}{1+\alpha} B + \mathcal{O}\left(\frac{1}{(1+\alpha)^2}\right)$$

$$N(\alpha) = \frac{\alpha}{1+\alpha} B + \mathcal{O}\left(\frac{1}{(1+\alpha)^2}\right)$$

for large $\alpha$ (Manteuffel, [6]). A quick calculation yields

$$M^{-1}(\alpha) = I + \frac{1}{1+\alpha} B + \mathcal{O}\left(\frac{1}{(1+\alpha)^2}\right).$$

This yields

$$M^{-1}(\alpha)A = I - B + \frac{1}{1+\alpha} (B-B^2) + \mathcal{O}\left(\frac{1}{(1+\alpha)^2}\right).$$

Asymptotically, this is the Jacobi splitting. Suppose $\mu_i$ is an eigenvalue of $A = I - B$ and $\mu_i(\alpha)$ is the corresponding eigenvalue of $M^{-1}(\alpha)A$. Then, for $\alpha$ sufficiently large we have

$$1 < \mu_i(\alpha) < \mu_i, \text{ for } \mu_i > 1 ,$$

and

$$\mu_i < \mu_i(\alpha) < 1, \text{ for } \mu_i < 1 .$$

Thus the condition of $M^{-1}(\alpha)A$ is less than the condition of the Jacobi splitting. In fact, one can show that for sufficiently large $\alpha$ we have the following:

$$\frac{\mu_i(\alpha)}{\mu_j(\alpha)} < \frac{\mu_i}{\mu_j} \quad \text{for } \mu_i \geq \mu_j .$$

This gives the following result (Manteuffel, [6]).

Theorem: For $\alpha$ sufficiently large, the shifted incomplete Cholesky factorization is better than the Jacobi splitting for acceleration by the generalized conjugate gradient iteration.

5. Choosing $\alpha$. The behavior of the splitting for small values of $\alpha$ is of greater interest, but not well understood. Suppose we are given the matrix A and the graph of L. Let $\alpha_a$ be such that the incomplete factorization is stable for every $\alpha > \alpha_a$. In general $\alpha_a$ is not known. Bounds on $\alpha_a$ may sometimes be found by examining the problem from which the linear system arose. We do have the following results (Manteuffel, [6]).

Corollary: If A is diagonally dominant or Stieltjes then $\alpha_a \leq o$.

Let $\alpha_c$ be the value of $\alpha$ that yields the minimum condition for $M^{-1}(\alpha)A$. Again, this is not known in general. We have the following.

Theorem: If A is a Stieltjes matrix and if the graph of L includes the graph of A, then if $o \leq \alpha_1 < \alpha_2$ the condition of $M^{-1}(\alpha_1)A$ is smaller than the condition of $M^{-1}(\alpha_2)A$ (up to a factor of 2).

Corollary: If A is a Stieltjes matrix, then $\alpha_c \leq o$ (up to a factor of 2).

Now consider the splitting that results from using

extend graph $G_i$. Let $\alpha_a{}^i$, $\alpha_c{}^i$, be the minimum $\alpha$ and $\alpha$ of best condition respectively for the splitting with graph $G_i$. In practice, one finds

$$\alpha_a{}^{i+1} \leq \alpha_a{}^i$$

and

$$\alpha_c{}^{i+1} \leq \alpha_c{}^i$$

however, this has not been proven. Let

$$c^i = C(M^{-1}(\alpha_c^i)A)$$

be the condition of the splitting using graph $G_i$ and $\alpha_c{}^i$. In practice one finds

$$c^{i+1} \leq c^i .$$

Again this has not been proven except in the case of Stieltjes matrices. A more important question is whether the improved condition justifies the additional work and storage required to perform the factorization with a larger graph and thus more fill-in.

7. Test Problem. In order to better understand the behavior of the splitting let us examine a test problem (see Manteuffel, [6]). Figure 3 shows a finite element mesh of a tapered wedge. Using isoparametric 20-node brick finite elements, structural deformation of the wedge was modeled. The boundary conditions and initial load corresponded to pressing the thin edge against a wall. This gave a linear system with 3090 unknowns, a half-bandwidth of 286 and a total of 170 000 nonzeroes in the lower triangular part of A. The condition of the matrix was on the order of $10^{+7}$. Figure 4 shows a comparison

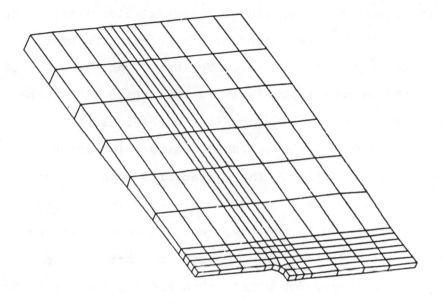

Figure 3.   Tapered Wedge

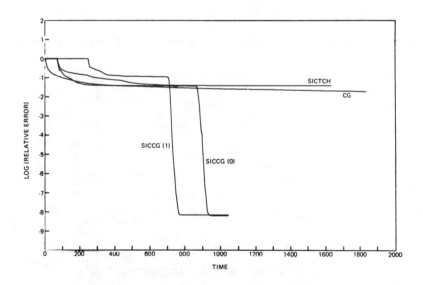

Figure 4.   Method Comparison

of the log of the Euclidean norm of the error versus cp
time in seconds on a CDC 6600 for several algorithms. The
competing methods were a conjugate gradient acceleration
(CG) of the Jacobi splitting, a Tchebychev acceleration of
the shifted incomplete Cholesky factorization with graph
$G_0$ (SICTCH), and conjugate gradient acceleration of the
shifted incomplete Cholesky factorization with graph $G_0$
and $G_1$ (SIC(0), SIC(1)). The values of $\alpha$ used were $\alpha =$
.005 for SIC(0) and $\alpha = .0015$ for SIC(1). The initial
plateau corresponds to the time required to perform the
factorization. Notice that although the factorization
time was much longer for SIC(1) than SIC(0), total time to
convergence was smaller for SIC(2). The steep cliff is
characteristic of the conjugate gradient iteration and may
be due to a bunching of the eigenvalues of $M^{-1}(\alpha)A$.

The number of iterative steps required to reach the
cliff was dependent upon the paramter $\alpha$. Figure 5 shows

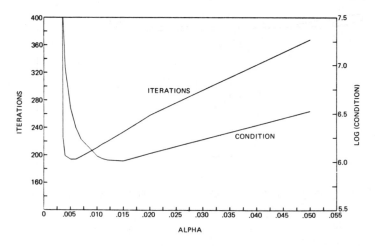

Figure 5. Iterations and Condition vs. $\alpha$

the number of iterations required to reduce the error by a
factor of $10^{-6}$ for various values of $\alpha$ using SIC(0).
Notice the unimodal shape of the graph.  It was found
experimentally that the factorization was unstable for
$\alpha < .0036$ and that convergence was fastest for $\alpha_b = .0055$.
Notice that for $\alpha = .05$, an order of magnitude larger,
convergence still occurred within a reasonable number of
iterations.

The condition of the splitting was estimated by taking
advantage of the relationship between the Lanczos algorithm
for finding eigenvalues and the conjugate gradient algorithm.
The log of the condition for various values of $\alpha$ is also
shown in Figure 5.  Notice that the best condition occurs
at $\alpha_c = .015$ and that

$$0 < \alpha_a < \alpha_b < a_c .$$

This was the case in every test.  Perhaps the value $\alpha_b$
produced a greater bunching of the eigenvalues which
outweighed the enhanced condition.

7.  Conclusion.  The incomplete Cholesky factorization
provides a bridge that spans the gap between direct and
iterative methods.  It appears that a factorization which
retains only a fraction of the fill-in yields an approximation that is sufficiently accurate on a large set of
vectors to effectively reduce the degree of the minimal
polynomial of the splitting.  The conjugate gradient
iteration exploits this situation.

The shifted factorization provides a means by which the incomplete Cholesky factorization can be extended to any positive definite linear system. The asymptotic behavior of the shift gives confidence that the factorization does yield improved condition. It does not, however, explain the remarkable performance for smaller values of $\alpha$. Hopefully, theory will be developed that will allow the a priori estimation of acceptable values for the shift parameter $\alpha$.

References.

1.  P. Concus, G. H. Golub, D. P. O'Leary, A Generalized Conjugate Gradient Method for the Numerical Solution of Elliptic Partial Differential Equations, Lawrence Berkeley Laboratory Pub. LBL-4604, Berkeley, CA.

2.  M. Engeli, T. Ginsburg, H. Rutishauser and E. Stiefel, Refined Iterative Methods for Computation of the Solution and the Eigenvalues of Self-Adjoint Boundary Value Problems, Birkhäuser Verlag, Basel/Stuttgart, 1959.

3.  Ky Fan, Note on M-matrices, Quart. J. Math., Oxford (2nd Series) 11 (1960), pp. 43-49.

4.  A. Greenbaum, Comparison of Splittings Used With The Conjugate Gradient Method, Lawrence Livermore Laboratories Report, UCRL-80800, March 1978.

5.  T. A. Manuteuffel, The Tchebychev Iteration for Nonsymmetric Linear Systems, Numer. Math., 28, 307-327 (1977).

6. T. A. Manteuffel, The Shifted Incomplete Cholesky Factorization, Sandia Report, SAND78-8226, Livermore, CA (1978).

7. J. A. Meijerink and H. A. van der Vorst, An Iterative Solution Method for Linear Systems of Which the Coefficient Matrix is a Symmetric M-matri, Math. Comp., vol. 31, 137 (1977), pg. 148-162.

8. R. S. Varga, Factorization and Normalized Iterative Methods, Boundary Problems in Differential Equations, R. E. Langer, Ed., University of Wisconsin Press, Madison (1960).

9. R. S. Varga, Matrix Iterative Analysis, Prentice Hall, Englewood Cliffs, New Jersey (1962).

# Algorithmic Aspects
# of the Multi-Level Solution
# of Finite Element Equations

## Randolph E. Bank* and A. H. Sherman*

Abstract. We discuss algorithmic aspects of subroutine
PLTMG, a Fortran program for solving self-adjoint elliptic
partial differential equations in general regions of $R^2$.
PLTMG is based on a piecewise-linear-triangle finite
element method, an adaptive grid refinement procedure, and
a multi-level iterative method to solve the resulting sets
of linear equations.

1. Introduction. Consider the model elliptic boundary
value problem

(1.1)
$$Lu = - \nabla \cdot (a \nabla u) + bu = f \text{ in } \Omega \subset R^2$$

$$u = g_1 \qquad \text{on } \partial\Omega_1$$

$$\frac{\partial u}{\partial n} = g_2 \qquad \text{on } \partial\Omega_2 = \partial\Omega - \partial\Omega_1$$

where the coefficient $a(x,y)$ $(b(x,y))$ is positive (nonnegative) in $\Omega$. In this work we discuss algorithmic

aspects of solving (1.1) using a Rayleigh-Ritz-Galerkin

method based on piecewise-linear triangular finite elements,

a multi-level iterative scheme for solving the resulting

matrix equations, and an adaptive grid refinement procedure.

For expositional convenience, we assume $g_1 = g_2 = 0$ and

that $\Omega$ is a polygon, although our FORTRAN subroutine PLTMG

is designed for the more general equation (1.1).

In the Rayleigh-Ritz-Galerkin procedure [10,11], we seek

---

*Dept. of Computer Science, University of Texas at Austin,
Austin, TX  78712

an approximate solution to the weak form of (1.1): find
$u \in H_E^1(\Omega)$ satisfying

(1.2)     $a(u,v) = (f,v)$ for all $v \in H_E^1(\Omega)$

where

$$a(u,v) = \int_\Omega a \, \nabla u \cdot \nabla v + buv \, dx \, ,$$

and $(\cdot, \cdot)$ denotes the usual $L^2(\Omega)$ inner product. We use
$H_E^1(\Omega) \subseteq H^1(\Omega)$ to denote the subspace of the usual Sobolev
space $H^1(\Omega)$ whose elements satisfy essential boundary
conditions [10,11]. Associated with the bilinear form
$a(\cdot, \cdot)$ is the energy norm $|||u|||^2 = a(u,u)$.

Let $T$ denote a triangulation of $\Omega$, and let $M \subset H_E^1(\Omega)$
denote the N-dimensional space of $C^0$ piecewise-linear
polynomials associated with $T$. The finite element approximation of u in (1.2) is the function $\tilde{u} \in M$ which satisfies

(1.3)          $a(\tilde{u},v) = (f,v)$ for all $v \in M$.

Once a basis $\{\phi_i\}_{i=1}^N$ for $M$ has been selected (the nodal
basis [10,11] is usually chosen), (1.3) can be reformulated
as a system of linear equations

(1.4)                    $AU = F$

where $A_{ij} = a(\phi_j, \phi_i)$ and $F_i = (f, \phi_i)$. Usually N is large
and the stiffness matrix A is sparse.

Our multi-level procedure involves a sequence of nested
triangulations $T_j$, with $T_{j+1}$ nested in $T_j$, $j \geq 1$, and
corresponding $N_j$-dimensional subspaces, $M_j$, of $C^0$ piecewise-linear polynomials. (By nested, we mean that each
triangle of $T_{j+1}$ intersects the interior of exactly one

triangle of $T_j$.) Corresponding to each subspace $M_j$ is the problem $P_j$, the analog of (1.3): Find $u_j \varepsilon M_j$ satisfying

(1.5)       $a(u_j, v) = (f, v)$ for all $v \varepsilon M_j$.

The multi-level method also requires the solution of problems $P_j'$ of more general form: Find $z_j \varepsilon M_j$ satisfying

(1.6)       $a(z_j, v) = G(v)$ for all $v \varepsilon M_j$

where $G(v)$ is a linear functional defined for $v \varepsilon M_j$. Both (1.5) and (1.6) require the solution of a linear system involving a stiffness matrix like A.

The current interest in multi-level methods is primarily due to the fact that they are of optimal order in terms of computational complexity [3-5,8-9]. That is, under certain hypotheses on the continuous problem (1.1)-(1.2) and the triangulations $T_j$, the work required to compute an approximation $\tilde{u}_j \varepsilon M_j$ to $u_j \varepsilon M_j$ of (1.5), which satisfies

(1.7)                      $||| \tilde{u}_j - u ||| \leq C N_j^{-q}$

is proportional to $N_j$. Here $C = C(u, a, b, \Omega, T_1)$ and $q$ is the "correct" exponent provided by the standard finite element analysis [10,11]. Several of these theoretical hypotheses may not be satisfied by some problems which may be solved using PLTMG; nonetheless, the multi-level scheme appears empirically to work well even in these cases.

There are five major organizational blocks within PLTMG: subroutines GRID, MATRIX, RHS, MG, and ADAPT. GRID is used to construct a sequence of triangulations according to user specifications. ADAPT is used to construct triangulations

using an adaptive refinement procedure based on the ideas
of Babuska and Rheinboldt [1,2]. In both cases, the user
need only describe $\Omega$ using a minimal number of triangles.
Our approach to the refinement problem is adapted from
techniques used by Bank and Dupont in the solution of
certain non-linear parabolic problems, in which the grid
was periodically redefined in order to track the propagation
of fronts and singularities [6]. Section 2 contains a
discussion of our refinement procedure.

MATRIX and RHS are used to assemble the matrices and
right hand sides for problems (1.4). Aside from the interface with our data structures, these routines are basically
standard, and we will not describe them here. MG is the
driver for the multi-level iteration; some details of this
part of our implementation are found in Section 3.

Since multi-level schemes employ a sequence of triangulations, they are ideally suited for use in conjunction
with adaptive refinement procedures in which the computed
solution $\tilde{u}_{j-1}$, corresponding to $T_{j-1}$, is used to determine
the refinement pattern for $T_j$. Section 4 describes this
aspect of subroutine ADAPT.

Finally in Section 5, we briefly describe the user
interface of PLTMG and present some numerical results.

2. Grid Refinement. As noted in the Introduction, a
major portion of PLTMG is devoted to grid refinement. In
this section we describe the procedure used to refine the
user-supplied triangulation $T_0$ of $\Omega$, while in Section 4,

we discuss the adaptive refinement process.

Initially, the user of PLTMG supplies a coarse triangulation $T_0$ of $\Omega$ consisting of a small number of triangles $t_i$, $1 \le i \le maxt_0$ (cf. Figure 2.1a). Each triangle $t_i$ contains three vertices $v_i^j$, $1 \le j \le 3$, and three edges $\varepsilon_i^j$, $1 \le j \le 3$, with $\varepsilon_i^j$ opposite $v_i^j$ (cf. Figure 2.1b). It is convenient to assign global numbers to the vertices and edges in $T_0$, denoted by $v_k$, $1 \le k \le maxv_0$, and $e_k$, $1 \le k \le maxe_0$, respectively. Thus for $1 \le i \le maxt_0$ and $1 \le j \le 3$, $v_i^j = v_k$ for some k, $1 \le k \le maxv_0$, and $\varepsilon_i^j = e_\ell$ for some $\ell$, $1 \le \ell \le maxe_0$. Throughout this paper, we will view local designations (e.g. $v_i^j$) and global designations (e.g. $v_k$) as interchangeable names for a unique entity, and we will use whichever designation makes more sense in context.

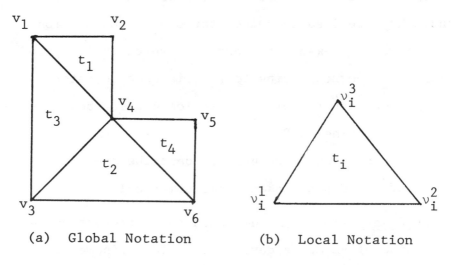

(a)  Global Notation          (b)  Local Notation

FIGURE 2.1

Each edge of a triangle $t_i$ either is a boundary edge of $\Omega$ or is part of the perimeter of one or more other triangles in the grid. We define the neighbor of $t_i$ across edge $\varepsilon_i^j$, denoted $\tau_i^j$, as the smallest regular triangle with one edge which completely overlaps $\varepsilon_i^j$. (A regular triangle is one obtained by regular subdivision; see below.) If $\varepsilon_i^j$ is a boundary edge, we define $\tau_i^j \leq 0$ (the value depending on the boundary conditions). Note that the neighbor relation need not be symmetric and is time-dependent.

Our refinement algorithm is motivated by three con-straints:

(i)   The size of the smallest interior angle of any triangle should be bounded away from 0;

(ii)   The transition between large and small triangles in the grid should be "smooth";

(iii)   The user of PLTMG should be able to control the amount of refinement in various regions of $\Omega$.

To ensure that (i) is met, we allow only two types of triangle subdivision: regular and "green".[*] In regular subdivision (cf. Figure 2.2a) a triangle $t_i$ is divided in-to four smaller triangles, denoted $t_{s_i+j}$, $0 \leq j \leq 3$, by joining the midpoints of its edges. Each of the four new triangles (called "sons of $t_i$") is similar to $t_i$ (its "father"), so that regular subdivision never reduces the size of the interior angles.

[*]The term "green" subdivision was used by Bank and Dupont and can be traced back to Donald Rose.

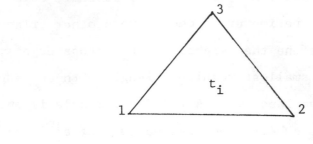

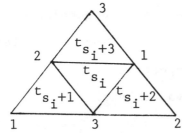 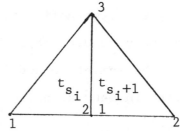

(a)  Regular Subdivision     (b)  "Green" Subdivision

(Vertex labels refer to the superscripts in $v_i^j$ notation.)

FIGURE 2.2

In green subdivision a triangle $t_i$ is divided into two smaller "green triangles", denoted $t_{s_i}$ and $t_{s_i+1}$, by inserting a "green edge" joining a vertex $v_i^j$ to the midpoint of the opposite edge $\varepsilon_i^j$ (cf. Figure 2.2b). Green subdivision may reduce the size of the smallest interior angle, so repeated use could violate (i). Hence we only use it to "clean up" the grid by removing degenerate quadrilaterals which remain after all regular subdivision has been completed.

To meet objective (ii), and to ensure that the "clean up" will involve only degenerate quadrilaterals, we divide a triangle $t_i$ using the regular subdivision process whenever two neighbors have been divided once, or one neighbor has been divided twice (cf. Figure 2.3).

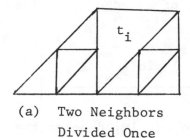 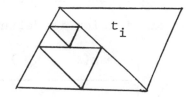

    (a)   Two Neighbors         (b)   One Neighbor
           Divided Once              Divided Twice

Situations Requiring Regular Subdivision of $t_i$

FIGURE 2.3

Finally, to allow the user of PLTMG to guide the refinement process, we introduce the notions of triangle and vertex level numbers. The level number of a triangle $t_i$, denoted $\ell_{t_i}$, is an indication of the number of subdivisions required to obtain $t_i$ from a triangle of $T_0$. Precisely, we define

$$\ell_{t_i} = \begin{cases} 1 & \text{if} & t_i \; \varepsilon \; T_0 \\ 1 + \ell_{f_i} & \text{otherwise}^* \end{cases}$$

The level number of a vertex $v_k$, denoted $\ell_{v_k}$, is used to control the refinement process in the sense that any triangle containing $v_k$ must be regularly subdivided if its level is smaller than $\ell_{v_k}$. The vertex levels for vertices in $T_0$ are user-specified, and through them the user of

*Here we use $f_i$ to denote the father of $t_i$.

PLTMG can cause different amounts of refinement to occur in different regions of $\Omega$. The vertex level of a vertex $v_k$ which is created at the midpoint of an edge $\varepsilon_i^j$ during regular subdivision is defined by

$$(2.1) \quad \ell_{v_k} = \left\lfloor \frac{q \cdot \max(\ell_{v_a}, \ell_{v_b}) + (100-q) \cdot \min(\ell_{v_a}, \ell_{v_b})}{100} \right\rfloor$$

where $v_a$ and $v_b$ are the endpoints of $\varepsilon_i^j$, $\lfloor x \rfloor$ denotes the largest integer less than or equal to x, and q is a user-specified parameter. By changing the parameter q, the user can adjust the sharpness of the grading of the mesh near vertices of large level number.

We can now give a high-level version of our refinement procedure in Algorithm 2.1. The key to an efficient implementation of this algorithm is a judicious choice of data structures to represent the triangulation. If we view $\Omega$ itself as a "pseudo-triangle" $t_0$ that is the father of each of the triangles in $T_0$, then it can be seen that the subdivision process generates a "triangle tree" in which the root is $t_0$, every other node is a triangle either in $T_0$ or created during refinement, and edges lead downward from a triangle to its sons. All internal nodes of the tree have exactly four sons, except that $t_0$ has $\text{maxt}_0$ sons and fathers of green triangles have two sons. The level number $\ell_{t_i}$ of a triangle $t_i$ now represents the length of the path from $t_i$ back to $t_0$ in the tree.

## Algorithm 2.1

Procedure Refine

    [$i \leftarrow 1$;   maxt $\leftarrow$ maxt$_0$ ;

    While ($i \leq$ maxt) do

       [For $j \leftarrow 1$ to 3 do

          [If $\tau_i^j$ is undivided Then

            If $\tau_i^j$ has two divided neighbors Then Divide($\tau_i^j$);

            Else If $\ell_{t_i} > \ell_{\tau_i^j} + 1$ Then Divide($\tau_i^j$)];

      If Dvtest($t_i$) Then Divide($t_i$);

      $i \leftarrow i + 1$];

    imax $\leftarrow$ maxt;

    For $i \leftarrow 1$ to imax do

      [If $t_i$ is undivided Then

        If $t_i$ has a divided neighbor Then Green($t_i$)]];

Procedure Divide($t_i$)

    [$s_i \leftarrow$ maxt + 1;

    maxt $\leftarrow$ maxt + 4;

    Create $t_{s_i+j}$, $0 \leq j \leq 3$, along with associated vertices];

Procedure Green($t_i$)

    [$s_i \leftarrow$ maxt + 1;

    maxt $\leftarrow$ maxt + 2;

    Create $t_{s_i}$ and $t_{s_i+1}$];

Procedure Dvtest($t_i$)

    [Dvtest $\leftarrow$ false;

    If $\max_{1 \leq j \leq 3} \ell_{v_i^j} > \ell_{t_i}$ Then Dvtest $\leftarrow$ true];

In PLTMG we represent the triangle tree with three arrays: a $3 \times$ maxt array (ITNODE) and two vectors of length maxt (F and S). In general, for a triangle $t_i$ we define

$$\text{ITNODE } (j,i) = k, \quad \text{where } v_i^j = v_k$$

$$F(i) = k, \quad \text{where } t_k \text{ is the father of } t_i$$

$$S(i) = \begin{cases} s_i & \text{if } t_i \text{ is divided} \\ 0 & \text{otherwise} \end{cases}$$

In order to distinguish a green triangle $t_i$, we negate $F(i)$ and $S(F(i))$. Figure 2.4 shows a simple three-level refinement process, the associated triangle tree, and the arrays ITNODE, F, and S.

In addition to the structure of the triangle tree, it is necessary to have access to $\ell_{t_i}$ and $\tau_i^j$, $1 \le j \le 3$, for each triangle $t_i$. In PLTMG, the triangle level numbers are stored in a vector L of length maxt with $L(i) = \ell_{t_i}$. In the case of $\tau_i^j$, $1 \le j \le 3$, however, we trade storage reduction for an increase in time. We compute and save $\tau_i^j$, $1 \le j \le 3$, for triangles $t_i \varepsilon \, T_0$, but for other triangles, compute this information by searching the triangle tree. The time required to find $\tau_i^j$ for a given $t_i$ is then proportional to the length of the path connecting $t_i$ and $\tau_i^j$.

To conclude this section we discuss the way that Algorithm 2.1 fits into the multi-level solution scheme. Applying the algorithm to the initial triangulation causes a sequence of regular triangle subdivisions to occur, eventually leading to a fully refined grid. However, we can stop the process early by limiting the maximum allowable triangle level number and, after adding necessary green edges, obtain a partially refined triangulation suitable for use with the multi-level scheme. To obtain

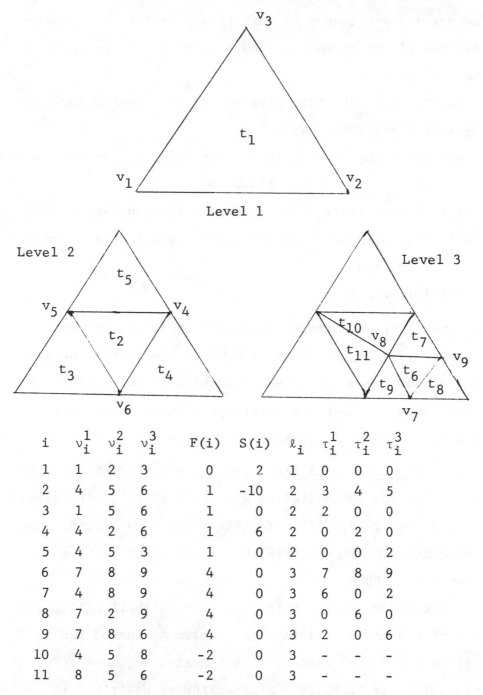

| i | $\nu_i^1$ | $\nu_i^2$ | $\nu_i^3$ | F(i) | S(i) | $\ell_i$ | $\tau_i^1$ | $\tau_i^2$ | $\tau_i^3$ |
|----|----|----|----|----|-----|----|----|----|----|
| 1 | 1 | 2 | 3 | 0 | 2 | 1 | 0 | 0 | 0 |
| 2 | 4 | 5 | 6 | 1 | -10 | 2 | 3 | 4 | 5 |
| 3 | 1 | 5 | 6 | 1 | 0 | 2 | 2 | 0 | 0 |
| 4 | 4 | 2 | 6 | 1 | 6 | 2 | 0 | 2 | 0 |
| 5 | 4 | 5 | 3 | 1 | 0 | 2 | 0 | 0 | 2 |
| 6 | 7 | 8 | 9 | 4 | 0 | 3 | 7 | 8 | 9 |
| 7 | 4 | 8 | 9 | 4 | 0 | 3 | 6 | 0 | 2 |
| 8 | 7 | 2 | 9 | 4 | 0 | 3 | 0 | 6 | 0 |
| 9 | 7 | 8 | 6 | 4 | 0 | 3 | 2 | 0 | 6 |
| 10 | 4 | 5 | 8 | -2 | 0 | 3 | - | - | - |
| 11 | 8 | 5 | 6 | -2 | 0 | 3 | - | - | - |

Triangle Tree Data Structures

FIGURE 2.4

the necessary sequence of such triangulations, we simply make use of an increasing sequence of limits on the triangle level numbers.

Algorithmically, this strategy is implemented for each limit in the sequence by

(i) replacing, for $v_k$ in $T_0$, the value $\ell_{v_k}$ by the smaller of $\ell_{v_k}$ and the desired bound;

(ii) recomputing $\ell_{v_k}$ for $v_k$ not in $T_0$ according to (2.1);

(iii) logically removing any green edges in the triangulation; and

(iv) invoking Refine.

The result is an efficient procedure that generates a sequence of triangulations of $\Omega$ which are nested (except for certain green triangles) and which satisfy the other constraints requisite to their use with the multi-level solution scheme (cf. Bank and Dupont [4]).

3. <u>The Multi-Level Iteration</u>. The term "multi-level iteration" can be applied to a diverse class of iterative methods (cf. [3-5,8-9]). In this section we describe the particular algorithm within this class which we have implemented in PLTMG.

Our algorithm is built upon what we shall call the k-level scheme for solving the problem $P'_k$ introduced in Section 1. Assume that we have a nested sequence of triangulations $T_1, T_2, \ldots, T_k$ generated as described in Section 2, and let $b_k(\cdot, \cdot)$ be a symmetric bilinear form associated with $T_k$. One step of the k-level scheme takes an initial approximation $z_0 \varepsilon M_k$ to a final approximation

$z_{m+1} \varepsilon M_k$, where m is a parameter of the method which may

be specified. This single step may be defined recursively

as follows:

(i) For $\ell = 1, 2, \ldots, m$, solve the following for $z_\ell$:

(3.1)  $b_k(z_\ell - z_{\ell-1}, \phi) = G(\phi) - a(z_{\ell-1}, \phi)$ for all $\phi \varepsilon M_k$,

(ii) Find $\tilde{\delta} \varepsilon M_{k-1}$, an approximation to $\delta \varepsilon M_{k-1}$,
        the solution of

(3.2)  $a(\delta, \phi) = G(\phi) - a(z_m, \phi) \equiv \overline{G}(\phi)$ for all $\phi \varepsilon M_{k-1}$,

(iii) Set

(3.3)  $$z_{m+1} = z_m + \tilde{\delta}.$$

Iteration (3.1) is called a smoothing iteration; its

purpose is to damp components of the error which oscillate

on the order of the mesh size in $T_k$. The bilinear form

$b_k(\cdot, \cdot)$ can represent any iterative method [12,13] which

will do this, although underrelaxed Jacobi-like schemes

are typically chosen in theoretical studies. In PLTMG,

however, we have chosen to implement a symmetric Gauss-

Seidel iteration (i.e., SSOR with $\omega = 1$) [12,13].

If the smoothing iteration is successful, then the error

becomes less oscillatory; and we can anticipate that it can

be well-approximated by an element in the space of smaller

dimension $M_{k-1}$. Thus in (3.2) we compute $\tilde{\delta}$, an approximate

elliptic projection of the error into $M_{k-1}$, using sparse

Gaussian elimination for k = 2 or two iterations of the

k-1 level scheme with initial approximation zero for k > 2.

In (3.3) we add the approximate error to $z_m$ to obtain

$z_{m+1}$. If $T_k$ is nested in $T_{k-1}$ (i.e. no triangles in $T_k$ intersect more than one triangle in $T_{k-1}$), then $M_{k-1} \subseteq M_k$ so that both $\tilde{\delta}$ and $z_{m+1}$ will be in $M_k$. In practice, however, this may not quite hold because some triangles which are divided into two green triangles in $T_{k-1}$ are regularly divided in $T_k$. In this event, we simply replace $\tilde{\delta}$ in (3.3) by its interpolant in $M_k$.

Using the k-level scheme (3.1)-(3.3) as an inner iteration, we can now build the overall multi-level iterative process aimed at solving the sequence of problems $P_j$, $1 \leq j \leq k$, introduced in Section 1. In this process we successively compute the approximate solutions $\tilde{u}_j$ of $P_j$, $1 \leq j \leq k$ as follows:

(i)   For j = 1, compute $\tilde{u}_1$ by solving (1.5) directly.

(ii)  For $2 \leq j \leq k$, compute $\tilde{u}_j$ using r iterations of the j-level scheme applied to (1.5) with $\tilde{u}_{j-1}$ as the initial approximation. (Here r is a parameter that may be user-specified.)

As an example, the multi-level process for the case k=3 is summarized below,

(i)   Solve directly for $\tilde{u}_1$

(ii)  Starting from $\tilde{u}_1$, carry out r iterations (a)-(c):

(a) m smoothing iterations at level 2

   (b) direct solution at level 1

(c) update iterate at level 2

take level 2 iterate as $\tilde{u}_2$

(iii) Starting from $\tilde{u}_2$ carry out r iterations of (a)-(h):

    (a) m smoothing iterations at level 3

      (b) m smoothing iterations at level 2

        (c) direct solution at level 1

      (d) update iterate at level 2

      (e) m smoothing iterations at level 2

        (f) direct solution at level 1

      (g) update iterate at level 2

    (h) update iterate at level 3

    Take level 3 iterate as $\tilde{u}_3$.

In (iii), the smoothing iterations at level 3 correspond to the original problem (1.5) with j = 3, those at level 2 correspond to the residual equations, and the direct solutions at level 1 correspond to the residual equations of the residual equations.

    To implement the multi-level scheme in general, we need four basic pieces of software in addition to a straightforward driver:

  (i)   a module to carry out m smoothing iterations for any problem $P'_j$, $j \geq 2$;

  (ii)  a module to compute the righthand side of (3.2) for any pair of spaces $(M_{j-1}, M_j)$, $j \geq 2$;

 (iii)  a module to carry out (3.3); and

  (iv)  a module to directly solve $P'_1$.

In PLTMG we have written our own subroutines for (i)-(iii) and made use of the Yale Sparse Matrix Package [7] for (iv).

4. Adaptive Refinement. In the multi-level scheme described in the previous section, no information about the level j grid $T_j$ is required until the approximation $\tilde{u}_{j-1}$ has been computed. Although we have assumed thus far that all grids are known in advance, it is clear that we may, in fact, use $\tilde{u}_{j-1}$ to adaptively determine $T_j$.

Our adaptive refinement procedure is based on the ideas of Babuska and Rheinboldt [1,2]. Let $t_i \varepsilon\, T_{j-1}$ have diameter $h_{t_i}$, and let

$$(4.1) \qquad |||z|||^2_{t_i} = \int_{t_i} a|\nabla z|^2 + bz^2 dx$$

denote the energy norm associated with $t_i$. We estimate the error, in energy, in $t_i$ using

$$(4.2) \quad |||\tilde{u}_{j-1}-u|||^2_{t_i} \cong C_1(a,t_i)h^2_{t_i}\int_{t_i} (L\tilde{u}_{j-1}-f)^2 dx$$

$$+ C_2(a,t_i)h_{t_i}\int_{\partial t_i} J^2 dx$$

where J is the jump in normal derivative in $\tilde{u}_{j-1}$ across $\partial t_i$. $C_1$ and $C_2$ are computable constants which depend on the coefficient function a(x,y) and the geometry of $t_i$ (but is independent of the size of $t_i$). The factors of $h_{t_i}$ are required to make the homogeneity of the right and left hand sides of (4.2) consistent. Formula (4.2) is modified slightly for triangles with one or more boundary edges. The right hand side of (4.2) is thus computable for each $t_i \varepsilon\, T_{j-1}$ using only $\tilde{u}_{j-1}$ and the partial differential equation.

We now describe the adaptive computation of $T_j$ from

$T_{j-1}$. Let $e_{max}$ denote the largest estimated error in any triangle in $T_{j-1}$, and let $t_{max} \varepsilon T_{j-1}$ be a triangle with estimated error $e_{max}$. Let $e_f$ denote the estimated error in the father of $t_{max}$, assuming it is known from a previous step of the adaptive procedure. If the local rate of convergence behaves like $Ch_{t_{max}}^q$ for some constants C and q, then the regular refinement of $t_{max}$ will result in estimated errors of approximate size $e_s \mathrel{\tilde{=}} e_{max}^2/e_f$ in the sons of $t_{max}$.

Therefore, to construct $T_j$ from $T_{j-1}$, we should divide only those triangles in $T_{j-1}$ which have estimated errors larger than the threshold value $e_{max}^2/e_f$. (If $e_f$ is not available, we assume q = 1, and set the threshold to $e_{max}/2$.) This procedure is easily integrated into the refinement process; all that is required is that we re-place procedure Dvtest described in Algorithm 2.1 with the procedure described above.

To save both time and space, we have actually imple-mented the adaptive scheme in PLTMG with two modifications:

(i) A triangle $t_i$ may be in both $T_{j-1}$ and $T_j$ if its error estimate is less than the threshold value. For such a triangle we do not recompute the error using $\tilde{u}_j$, but instead use the estimate computed previously using $\tilde{u}_{j-1}$.

(ii) If there is a severe singularity, it is possible that only a few triangles near that singularity will have errors larger than the threshold and be refined. If this were to occur in several consecutive refinements, the result would be a sequence of subspaces $M_j$ of slowly increasing dimension, requiring extra storage for unnecessary matrices, and extra work for the multi-level iteration. To avoid this, we have incorporated a "level compression" feature into PLTMG. If sufficiently few vertices have been added in an adaptive refinement step, we discard the old triangulation $T_{j-1}$ in favor of the newly created one. In this fashion, we generate a sequence of subspaces $M_j$ whose dimensions approximately satisfy

$$2N_{j-1} \leq N_j \leq 4N_{j-1},$$

ensuring a geometric increase in dimension.

We do not know theoretically what effect level compression has on the rate of convergence. Current convergence proofs explicitly or implicitly assume a sequence of quasi-uniform grids, which are not likely to be produced in situations where level compression is employed [3-4,8-9]. In practice, however, the procedure appears to have the desirable effects of reducing both work and storage, with only a modest reduction in the observed rate of convergence.

5. <u>User Interface/Numerical Results</u>. There are many possible ways to specify, in FORTRAN, a region in $R^2$, an elliptic operator, right hand side, and boundary conditions. In PLTMG, the user specifies these through the calling sequence

(CALL PLTMG (VX, VY, LXY, XM, YM, ITNODE, ITEDGE,

AXY, BXY, FXY, GXY, IP, W).

VX and VY are real arrays of length $maxv_0$, containing x
and y coordinates of the vertices in $T_0$. LXY is an integer array of length $maxv_0$ whose j-th entry is the level
number of vertex $v_j = (VX(j), VY(j))$. ITNODE is a 3 x $maxt_0$
array containing vertex numbers $v_i^j$ for $t_i \in T_0$. Triangles
on the boundary of $\Omega$ may have one curved boundary edge;
let $maxc_0 - 1$ denote the number of curved edges. XM and YM
are real arrays of length $maxc_0$ whose k-th entries contain
x and y coordinates of the midpoint of the k-1-st   curved
edge (the first entries are arbitrary).* ITEDGE is a
3x$maxt_0$ integer array, where ITEDGE (j,i) contains information about $\varepsilon_i^j$:

ITEDGE (j,i) =

$\quad$ -1 if $\varepsilon_i^j$ is a straight Dirichlet
$\qquad$ boundary edge

$\quad$ -k if $\varepsilon_i^j$ is a curved boundary edge
$\qquad$ with midpoint (XM(k),YM(k))

$\quad$ 0  if $\varepsilon_i^j$ is an interior edge

$\quad$ 1  if $\varepsilon_i^j$ is a straight Neumann
$\qquad$ boundary edge

$\quad$ k  if $\varepsilon_i^j$ is a curved Neumann boundary
$\qquad$ edge with midpoint
$\qquad$ (XM(k),YM(k))

$\quad$ AXY, BXY, FXY, and GXY are user supplied functions
which define the elliptic partial differential equations
(1.1). AXY, BXY and FXY have arguments (x,y,i) where
$(x,y) \in t_i$, $t_i \in T_0$. GXY has arguments (x,y,i,j), where
$(x,y) \in \partial\Omega$, $(x,y) \in \varepsilon_i^j$, $t_i \in T_0$. IP is an integer array of

*PLTMG approximates a curved edge by the arc of the circle
passing through the endpoints of the edges and the midpoint.

length 100. In some entries, the user specifies input parameters ($maxv_0$, $maxt_0$, $maxc_0$, m, r, etc.) and switches indicating the tasks to be performed; other entries contain
internally generated parameters, error flags, and pointers
to the work array W, from which internal storage is
allocated.

Output routines are available to carry out the
following tasks:

(i)    evaluate the solution and/or gradient at a list
of user specified points;

(ii)    estimate the convergence of the sequence $u_j$ to u,
measured in the energy, $L^2(\Omega)$, and $L^\infty(\Omega)$ norms;

(iii)    study the rate of convergence and computational
work for the multilevel iteration as a function of
m and r;

(iv)    plot the triangulations $T_j$ and/or draw contour
plots of the solution and/or the magnitude of the
gradient of the solution;

(v)    print the triangle tree data structures;

(vi)    compute the error in the energy norm on a tri-
angle-by-triangle basis in the finest mesh for
comparison with estimate (4.2).

(Note (ii) and (vi) above require the user to provide
the exact solution through a function UXY.) Software for
(iv) was written by C. Durrin.

To illustrate the behavior of PLTMG, we consider the
problem

(5.1)                    $-\Delta u = 0$                in $\Omega$

$u = r^{\frac{1}{2}}\sin(\theta/2)$ on $\partial\Omega^*$

where $\Omega$ is the square $|x| < 1$, $|y| < 1$ with a crack at
$0 \leq x \leq 1$, $y = 0$. The exact solution, $u = r^{\frac{1}{2}} \sin(\theta/2)$,
displays the principal part of the singularity due to the
crack [11]. We used an initial triangulation, $T_0$,
consisting of eight triangles and ten vertices as
illustrated in Figure 5.1.

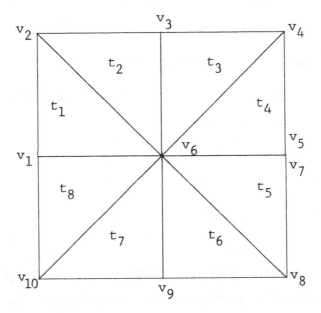

$v_5$ and $v_7$ have the same $(x,y)$ coordinates but $v_5$ is
above the crack and $v_7$ below.

FIGURE 5.1

We solved (5.1) twice, each time using four multi-grid
levels; in each case, $T_1 = T_0$. In the first case,

*For convenience in exposition we use polar coordinates
to represent the boundary conditions.

$T_j$, $2 \leq j \leq 4$, was constructed by regular refinement of each triangle in $T_{j-1}$. In the second case, triangulations were generated by the adaptive procedure using level compression. The smallest triangles in the adaptive case were located near the crack tip and were at level 7 in the triangle tree. These were smaller in size, by a factor of 64, than the smallest triangles in the uniform case, which were at level 4 in the tree (see Figure 5.2).

In Table 5.1, we indicate the rate of convergence of the sequence $u_j$ to u, where

(5.2)     $\text{Digits} = -\log_{10}\{ |||u_j - u||| / |||u||| \}$.

(The $C^0$ piecewise quadratic interpolant of u with respect to the finest grid was used for purposes of computing the error.) Using this data, we did a least squares fit of the data to an error bound of the form

(5.3)

$$\frac{|||u_j - u|||}{|||u|||} \cong CN_j^{-q/2}$$

to obtain C and q. Because of the singularity in u, the optimal value of q as $N \rightarrow \infty$ for uniform grids is $q = \frac{1}{2}$ [1,11]; the best possible rate of convergence for piecewise linear finite elements in general is $q = 1$, a rate which was essentially achieved by the adaptive procedure.

We next did a convergence study to determine the rate of convergence of $\tilde{u}_4$ to $u_4$ as a function of m and r, where

(5.4)   $\text{Digits} = -\log_{10}\{ |||\tilde{u}_4 - u_4||| / |||u_4||| \}$.

The results appear in Table 5.2.

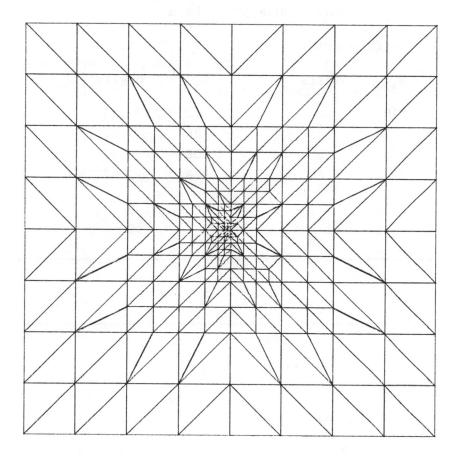

FIGURE 5.2

Table 5.1

Convergence of $\tilde{u}_j$ to u

| uniform grids | | adaptive grids | |
|---|---|---|---|
| $N_j$ | digits | $N_j$ | digits |
| 10 | .23 | 10 | .23 |
| 27 | .39 | 36 | .47 |
| 85 | .53 | 108 | .67 |
| 297 | .68 | 219 | .92 |

$$c = 1.2 \qquad\qquad c = 1.9$$
$$q = .61 \qquad\qquad q = .99$$

Table 5.2

Convergence of $\tilde{u}_4$ to $u_4$

| uniform grids (digits) | | | | | adaptive grids (digits) | | | |
|---|---|---|---|---|---|---|---|---|
| r/m | 1 | 2 | 3 | 4 | r/m | 1 | 2 | 3 | 4 |
| 1 | 1.1 | 1.5 | 1.7 | 1.8 | 1 | .9 | 1.1 | 1.2 | 1.3 |
| 2 | 1.7 | 2.3 | 2.6 | 2.8 | 2 | 1.2 | 1.5 | 1:7 | 1.9 |
| 3 | 2.2 | 3.0 | 3.5 | 3.9 | 3 | 1.4 | 1.8 | 2.1 | 2.4 |
| 4 | 2.7 | 3.8 | 4.4 | 4.8 | 4 | 1.6 | 2.1 | 2.5 | 2.9 |

For the case of quasi-uniform grids, Bank and Dupont [4] have shown that the energy norm of the error is reduced by a factor of at least $(Cm^{-\alpha})^r$, where C and $\alpha$ are postive constants independent of j, but which depend on the geometry and sizes of triangles $T_1$ and the partial

differential equation.  Technically, the adaptive case can
be forced into this framework as well, since we are
dealing with a fixed, finite number of multi-grid levels.
The slower rate of convergence to $u_4$ for the adaptive
grids appears to be due to a combination of the presence
of triangles of more widely varying sizes in the adaptive
grids (leading to matrix elements of more widely varying
magnitude), the use of level compression which accentuates
this variance, and the ordering of the vertices.  Note,
however, that in both cases, for the choice $m = r = 1$, $\tilde{u}_4$
already has essentially all the significant digits which
$u_4$ has as an approximation to the solution of the partial
differential equation.

In Table 5.3, we have listed the work estimates for
the computation of $\tilde{u}_4$ as a function of m and r.  Work is
measured in terms of matrix multiples in the finest grid,
estimated by

$$(5.5) \qquad \text{work} = c_1(c_2 m + 1)r + c_3 r(N_1 \log(N_1)/N_4).$$

Here $c_1$ depends on the ratios $N_j/N_4$, $1 \leq j \leq 3$, $c_2$ is the
number of matrix multiples per symmetric Gauss Seidel
iteration (about 1.5), and $c_3$ depends on the number of
multigrid levels and the fill-in generated in the sparse
factorization of the level one matrix.  The differences
in work estimates for the two problems are attributed
mainly to the fact that the ratio $N_3/N_4 \stackrel{\sim}{=} .5$ for the
adaptive grid, while for the uniform case it was closer to
.3.

Table 5.3

Work Estimates for Computing $\tilde{u}_4$

| | uniform grids | | | | | adaptive grids | | | |
|---|---|---|---|---|---|---|---|---|---|
| r/m | 1 | 2 | 3 | 4 | r/m | 1 | 2 | 3 | 4 |
| 1 | 6.2 | 9.9 | 13.5 | 17.2 | 1 | 9.1 | 14.6 | 20.0 | 25.5 |
| 2 | 12.3 | 19.7 | 27.1 | 34.4 | 2 | 18.3 | 29.2 | 40.1 | 51.0 |
| 3 | 18.5 | 29.6 | 40.6 | 51.6 | 3 | 27.4 | 43.7 | 60.1 | 76.4 |
| 4 | 24.7 | 39.4 | 54.1 | 68.8 | 4 | 36.5 | 58.3 | 80.1 | 101.9 |

Since, in general, for a k level problem PLTMG always has $N_j/N_k < L^{k-j}$, $L < 1$ and independent of j, $c_1$ can always be bounded independent of j. Furthermore since the second term (with $N_k$ replacing $N_4$) is usually small in comparison with the first, the work essentially depends linearly on m and r. As m becomes large, we derive less benefit from the additional smoothing iterations done on the coarser grids. If the bound $(Cm^{-\alpha})^r$ is sharp, one can show the optimal choice of m over a large class of interesting problems is bounded by a small number, e. g. 3-4.

## REFERENCES

1. I. Babuška and W.C. Rheinboldt, Error estimates for adaptive finite element computations, SIAM J. Numer. Anal., 15(1978), 736-754.

2. I. Babuška and W.C. Rheinboldt, Analysis of optimal finite element meshes in $R^1$ , Technical Note BN-869, Institute for Physical Science and Technology, University of Maryland, March, 1978.

3.  N.S. Bakhvalov,  On the convergence of a relaxation method with natural constraints on the elliptic operator, Zh. Vychisl. Mat. mat., 6(1966), 861-885.

4.  R.E. Bank and T. Dupont, An optimal order process for solving finite element equations, (revised January 1978), submitted.

5.  A Brandt, Multi-level adaptive solutions to boundary value problems, Math. of Comp., 31(1977), 333-390.

6.  J. Douglas, Jr. and T. Dupont, Interior penalty procedures for elliptic and parabolic galerkin methods, Lecture Notes in Physics #58, Springer-Verlag, 1976.

7.  S.C. Eisenstat, M.C. Gursky, M.H. Schultz, and A.H. Sherman, Yale sparse matrix package I:  the symmetric codes, Research Report #112, Department of Computer Science, Yale University, 1977.

8.  W. Hackbusch, On the convergence of a multi-grid iteration applied to finite element equations, Technical Report #77-78, Universitat Zu Koln, July, 1977.

9.  R.A. Nicolaides, On the $\ell^2$ convergence of an algorithm for solving finite element equations, Math. of Comp., 31(1977), 892-906.

10. J.T. Oden and J.N. Reddy, An Introduction to the Mathematical Theory of Finite Elements, J. Wiley and Sons, 1976.

11. G. Strang and G. Fix, An Analysis of the Finite Element Method, Prentice-Hall, 1973.

12. R.S. Varga, Matrix Iterative Analysis, Prentice-Hall, 1962.

13. D. Young, Iterative Solution of Large Linear Systems, Academic Press, 1971.

# LASCALA—
# A Language for Large Scale Linear Algebra

## A. W. Westerberg* and T. J. Berna*

**Abstract.** A problem-oriented language is described which is capable of directing the calculation sequence associated with solving large, sparse, linear algebra systems which in general will require the use of mass memory. Although the concept of LASCALA arose in connection with the problem of optimizing large chemical processes, its application is suitable for use with any large matrices having a loosely connected block diagonal structure.

1. _Introduction._ One approach used to develop a model of a chemical process is to write all of the constraints associated with the process and to solve them simultaneously. These constraints include linear and nonlinear equality and inequality algebraic constraints, and for a typical process there are often several thousand such constraints. Once the model has been generated, one would like to find the set of feasible (or optimal feasible) solutions which describe the process. Unfortunately, it is frequently undesirable, if not impossible, to handle all of the constraints simultaneously without resorting to the use of mass memory devices for auxiliary storage.

The purpose of our work has been to develop an optimization scheme capable of finding optimal feasible solutions for large chemical processes. Our first step was to develop a scheme for computing the L/U factors of the large Jacobian matrix associated with the process constraints. Our algorithm (Westerberg and Berna, 1978) performs the factorization in a block-by-block manner to avoid using an excessive amount of core storage. In order to implement the ideas presented in that paper and in order to extend these ideas for use with our optimization algorithm (Berna, Locke and Westerberg, 1978), we advocate

*Carnegie-Mellon University, Pittsburgh, Pennsylvania 15213. This work supported by NSF Grant ENG76-80149.

the use of a problem-oriented language such as LASCALA (Large Scale Linear Algebra).

We are in the early stages of developing LASCALA. Although the sample problem presented here has a chemical engineering parentage, LASCALA is well-suited for use in solving any problems that give rise to a large bordered block diagonal (BBD) Jacobian matrix. Presently LASCALA programs must be written manually, but its full potential can be realized only when the programs are generated automatically. Our primary objective in writing this paper is to introduce LASCALA in concept and to show the types of commands that such a language should have.

Throughout the discussion we assume that the user has available a sparse matrix package capable of performing an L/U factorization of a sparse matrix and of performing the forward and backward substitutions required to solve linear algebraic systems. We have given this hypothetical package the name SPARSE and assume it contains four routines: ANALYZE, FACTOR, FWD and BACK. The ANALYZE step, which determines the pivot sequence, must have some provision for handling nonpivot flags. FACTOR performs the elimination for a given pivot sequence. The routines FWD and BACK must be able to perform separately the forward and backward steps of the Gaussian elimination where the coefficient matrix is the factored matrix or its transparse. Our intention in designing LASCALA has been to keep the problem-oriented language independent of any particular sparse matrix code.

The discussion which follows is divided into six sections. The first section describes the nature of the chemical process design problem. Following a statement of the design problem, there is a statement of the optimization algorithm and a description of the method used to generate the process constraints. Section 5 describes some of the sparse matrix manipulations required in our work. In Section 6 we describe LASCALA in greater detail, and we discuss future work in Section 7.

2. The Design Process. A typical design problem is stated as follows: Design a process for producing 1 million kilograms of a chemical "C" per year. The product must be 99.9% pure, and available raw materials are chemicals "A" and "B". In addition to the explicit

requirements stated above, there are several implicit constraints requiring that the final design must: minimize annual operating expenses, comply with local pollution codes, meet certain safety standards, etc. A designer faced with solving this problem normally begins by considering each of the elements which might go into the process: the reactor, purification units, heat exchangers, etc. Before analyzing the behavior of the entire process the designer must analyze the behavior of each element. In considering the reactor, for example, the designer must consider various operating conditions available and alternative reaction schemes. As each of the elements is analyzed, the designer begins to link them together and to analyze the behavior of the entire network. The designer then continues to work with the network until an acceptable (perhaps an optimal) process is discovered.

A desirable feature of any design package is that the user must be able to place arbitrary specifications on the process. For example, the original problem statement might specify the temperature, pressure, composition or flowrate at any point in the process. The specifications might require that some function of these variables be satisfied. In any event, the final design must satisfy all specifications imposed on the process.

When the entire chemical process is considered there are $n+r+q$ variables and n equality constraints. The user specifies values for q of the variables; effectively, these variables become constants for the remainder of the problem. The Jacobian matrix for the equality constraints is factored into the product of a lower triangular matrix $\underline{\underline{L}}$ and an upper triangular matrix $\underline{\underline{U}}$. The n variables corresponding to the pivots of the factored Jacobian matrix become the dependent variables $\underline{x}$ and the remaining variables become the decision (independent) variables $\underline{u}$. The r decision variables represent the actual degrees of freedom for optimizing the chemical process design.

Before concluding our discussion of the chemical process design problem, we would like to say a few words about the structure of the Jacobian matrix corresponding to a typical chemical process. The Jacobian has a bordered block diagonal structure; the blocks correspond to the various elements in the process, and the overlap between blocks is due to the connections among the elements in the network. One can expect to find Jacobian matrices having this structure whenever the constraints are related to a set of loosely connected modules. We say more about the role that this structure plays in the design process later in this paper. Now let us consider the optimization problem as a mathematician might see it.

3. The Optimization Algorithm. The problem of finding an optimal design for a particular chemical process can be stated as follows:

$$\text{Min } \Phi(z)$$

subject to

(P1)
$$g(z) = 0$$
$$h(z) \geq 0$$
$$z \in R^{n+r}$$
$$g: R^{n+r} \to R^n$$
$$h: R^{n+r} \to R^m$$
$$\Phi: R^{n+r} \to R$$

where $\Phi$ may be the net annual operating expenses, and g and h are the constraints described in Section 2. For problems of industrial significance n and m range from about 1000 to 50,000 and r ranges from 1 to about 50. Typical values for n and r are 10,000 and 10, respectively. In order to solve (P1) within a reasonable amount of core storage and execution time, we have modified an algorithm published by Powell (1977). Powell's algorithm relies heavily on some work published by Han (1975). In the remainder of this section we sketch the development of the relevant details of our optimization algorithm (Berna, Locke and Westerberg, 1978).

In order to develop the conditions necessary for z* to be an optimal feasible solution to (P1), we define the Lagrange function

(1)
$$L(z,\lambda,\mu) = \Phi(z) - \lambda^T g(z) - \mu^T h(z)$$

The necessary conditions for optimality then become

(2)
$$\frac{\delta L}{\delta z} = 0 = \frac{\delta \Phi}{\delta z} - \frac{\delta g^T}{\delta z} \lambda - \frac{\delta h^T}{\delta z} \mu$$

(3)
$$g(z) = 0 \quad ; \quad h(z) \geq 0$$

(4)
$$\mu^T h = 0 \quad ; \quad \mu \geq 0$$

The Newton-Raphson scheme for solving (2) gives

(5)
$$\frac{\delta L}{\delta z} + \frac{\delta^2 L}{\delta z^2} \Delta z = 0 = \frac{\delta \Phi}{\delta z} + \frac{\delta^2 L}{\delta z^2} \Delta z - \frac{\delta g^T}{\delta z} \lambda - \frac{\delta h^T}{\delta z} \mu$$

In a similar fashion, the constraints in (3) can be linearized to give

$$g + \frac{\delta g}{\delta z^T} \Delta z = 0$$

(6)

$$h + \frac{\delta h}{\delta z^T} \Delta z \geq 0$$

At this point, we define the following quadratic function

(7)
$$Q(\Delta z) = \Phi + \frac{\delta \Phi}{\delta z^T} \Delta z + \frac{1}{2} \Delta z^T \frac{\delta^2 L}{\delta z \delta z^T} \Delta z$$

Now consider the quadratic programming problem (QPP) formed by minimizing $Q(\Delta z)$ subject to the linearized constraints (6):

$$\begin{array}{c} \text{Min } Q(\Delta z) \\ \Delta z \end{array}$$

(QPP)
$$\text{s.t.} \quad g + \frac{\delta g}{\delta z^T} \Delta z = 0$$

$$h + \frac{\delta h}{\delta z^T} \Delta z \geq 0$$

The necessary conditions for $\Delta z*$ to be a solution to (QPP) are that (5) and (6) hold and that the following constraints be satisfied:

(8)
$$\mu^T \left( h + \frac{\delta h}{\delta z^T} \Delta z* \right) = 0$$

$$\mu \geq 0$$

Based on the observation that (5), (6) and (8) are satisfied by solving (QPP) and that these constraints represent the Newton-Raphson iteration from $z_k$ to $z_{k+1}$, Powell (1977) developed an algorithm which solves (P1) by generating and solving a sequence of QPP's. Instead of actually computing $\dfrac{\delta^2 L}{\delta z \delta z^T}$, the Hessian of the Lagrange function, Powell uses a series of pairwise rank-one updates to approximate this matrix. The basic algorithm is given as follows:

Step 1.    Guess $C = I$ and $z$ $\left( C \text{ is } \dfrac{\delta^2 L}{\delta z \delta z^T} \right)$

Step 2.    Evaluate $\Phi$, $\dfrac{\delta \Phi}{\delta z}$, $g$, $h$, $\dfrac{\delta g}{\delta z^T}$, $\dfrac{\delta h}{\delta z^T}$, $\dfrac{\delta L}{\delta z}$

Step 3.    Solve QPP for $\Delta z*$, $\lambda*$, $\mu*$

Step 4.    Estimate $\left( \dfrac{\delta L}{\delta \Delta z} \right)_{z+\Delta z,\text{est}} = \left( \dfrac{\delta \Phi}{\delta z} \right)_z + \left( \dfrac{\delta^2 L}{\delta z \delta z^T} \right)_z \Delta z$

$$- \left( \dfrac{\delta g^T}{\delta z} \right)_z \lambda* - \left( \dfrac{\delta h^T}{\delta z} \right)_z \mu*$$

Step 5.    Move to $z_{\text{next}} = z + \Delta z$ and repeat Step 2. Using the same $\lambda*$ and $\mu*$ from Step 4, evaluate

$$\left( \dfrac{\delta L}{\delta \Delta z} \right)_{z_{\text{next,act}}} = \left( \dfrac{\delta \phi}{\delta z} \right)_{z_{\text{next}}} - \left( \dfrac{\delta g^T}{\delta z} \right)_{z_{\text{next}}} \lambda* - \left( \dfrac{\delta h^T}{\delta z} \right)_{z_{\text{next}}} \mu*$$

(At $z_{\text{next}}$, $\Delta z = 0$ so the term $\dfrac{\delta^2 L}{\delta z \delta z^T} \Delta z$ is zero.) Then use

$$\delta = \left[ \left( \dfrac{\delta L}{\delta \Delta z} \right)_{\text{est}} - \left( \dfrac{\delta L}{\delta \Delta z} \right)_{\text{act}} \right] \text{ to update } C.$$

Step 6.    Iterate from Step 3 until $||\Delta z||$ is small.

For relatively small problems ($n \leq 50$) this algorithm works extremely well. For very large problems ($n \geq 1000$) the size of $C$ exceeds the core storage space available on most machines; furthermore, the computational requirements involved in updating $C$ (by computing $W_k W_k^T$) are prohibitive for large values of $n$. Our algorithm, which extends the above algorithm, uses the linearized equality constraints in (6) to set up a <u>reduced</u> (QPP) where the size of the Hessian matrix is rxr instead of (n+r)x(n+r). A second advantage associated with the extended algorithm is that we never compute $W_k W_k^T$; instead, we always compute the appropriate scalar products associated with the pre- and post-multiplication operators of C. In other words, we never need C alone. Rather, we need terms of the form $p^T Cq$ where these terms are computed (after $\ell$ rank-one updates of C, $\ell \ll n$) as follows

$$p^T Cq = p^T q + \sum_{k=1}^{\ell} (p^T W_k)(W_k^T q)$$

This operation requires $(2\ell+1)$ n+$\ell$ multiplications; if C is treated as a full matrix, this operation requires $n^2$+n multiplications. This difference does not even include the $\ell n^2$ multiplications (for the $\ell$ outer products $W_k W_k^T$) required to compute C.

The basic approach of our algorithm is to partition the original variable set z into $x \in R^n$ and $u \in R^r$. This partitioning is accomplished by performing an L/U factorization of the matrix $\dfrac{\partial g}{\partial z^T}$. The variables corresponding to nonzero pivots are labeled x; the remaining variables are labeled u. At each iteration we use these factors to set up a reduced QPP which is solved to obtain $\Delta u$ and $\mu$. The values for $\Delta x$ and $\lambda$ are computed by performing the back substitution based on $\Delta u$ and $\mu$. Figure 1 illustrates this process. The details of the optimization algorithm are published elsewhere (Berna, Locke, Westerberg, 1978), but in the present discussion we wish to focus on the sparse matrix manipulations required to carry out the optimization procedure. Before discussing these manipulations we first need to describe the process for generating the Jacobian matrices and residuals associated with each of the constraints in the process. Section 4 describes this procedure, and we continue our discussion of the sparse matrix operations in Section 5.

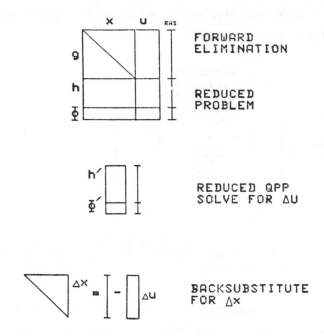

Figure 1.  Schematic diagram of procedure used to solve very large
            quadratic programming problems (QPP's).

4.  Process Model: Packets and Generators.  The discussion in
this section centers around the approach adopted for modeling a chemical process.  The job of the process model is to generate the constraints in (P1) that describe a process.  Conventional chemical process simulation packages visualize the process model as a set of interconnected unit subroutines which operate on inlet stream values
and produce outlet stream values.  In addition to the process streams
associated with each unit, there may be one or more parameters which
the user is required to specify.  As we mentioned earlier it is far
more desirable for a process simulator to accept arbitrary constraints
on the process; such is the case with equation based simulation packages.  We are interested in developing a process simulator capable
of generating the Jacobian elements and right-hand sides associated
with the constraints in (P1).  Jacobian elements are computed by
"generators" which compute the elements based on the values of variables in the associated "variable packets."  The rows are identified
as those belonging to the "equation packets" associated with the

particular generator. Each generator has a set of packets associated with it; Figure 2 illustrates.

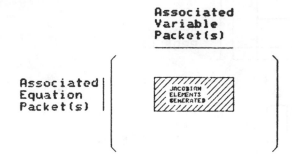

Figure 2.  Jacobian elements are generated by a generator in the
particular rows and columns identified as the associated
equation and variable packet(s).

In order to clarify the presentation of these concepts, we have developed a Generator, Equation and Variable (GEV) diagram which schematically illustrates the connection between the various generators and packets in a given process model. Figure 3 illustrates the GEV diagram and corresponding Jacobian matrix for a simple process. Variable packets are illustrated by labeled solid horizontal lines, equation packets by labeled solid vertical lines. Generators are represented by labeled boxes, and the packets associated with each generator are those connected by a dashed line to that generator. In this example variable packets "S2" and "Cost" are associated with two generators while the equation packet, "Cost Function," is associated with two generators. This diagram illustrates the fact that each generator may be associated with many packets and that any packet may be associated with more than one generator. From a chemical engineering viewpoint this concept in modeling offers several advantages: one is that the physical property calculations may be included as separate generators, another is that these physical properties can be associated with a process stream instead of using the less natural association of physical properties with process units. To those unfamiliar with chemical process models, this latter distinction may seem to be of little importance. The significance of the statement is more clearly understood when one considers that:  (1) all process

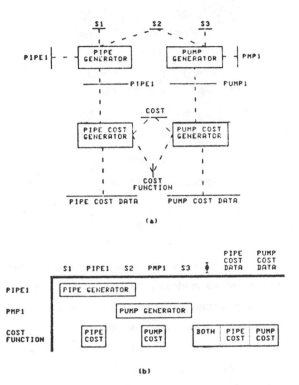

Figure 3. GEV diagram (a) for process containing a pump and a length of pipe; (b) shows Jacobian matrix associated with this GEV.

simulators currently associate physical property calculations with a process <u>unit</u> and not with a process <u>stream</u>, and (2) because a process stream is usually associated with two process units there is the possibility that two different values for a computed physical property could be assigned to the same process stream. Our convention can eliminate the possibility of having this discrepancy arise.

   5. <u>Manipulations for Large Problems</u>. In Section 3 we alluded to an optimization algorithm which solves (P1) by solving a series of QPP's in the degrees of freedom, $\Delta u$, only. In order to arrive at this reduced QPP one must be able to compute the L/U factors of the Jacobian matrix $\frac{\partial g}{\partial x^T}$. For large problems, those of industrial significance, the Jacobian matrix must be factored in a block-by-block manner; in this section we discuss the manipulations required to solve (P1) in a reasonable amount of core storage space. For convenience we introduce the following notation:

$$J_x = \frac{\partial g}{\partial x^T} \qquad J_u = \frac{\partial g}{\partial u^T}$$

$$A = J_x^{-1} J_u \qquad v = J_x^{-1} g$$

$$K_x = \frac{\partial h}{\partial x^T} \qquad K_u = \frac{\partial h}{\partial u^T}$$

$$\hat{C} = \text{Quasi-Newton Approximation to } \frac{\partial^2 L}{\partial z \partial z^T}$$

$$w_k = k^{th} \text{ rank-one update to } \hat{C}$$

Although the ideas presented in this section first arose in connection with chemical process models, we wish to stress that they may be applied to any loosely connected network of modules giving rise to a bordered block diagonal (BBD) Jacobian matrix.

The structure of a typical chemical process naturally gives rise to a bordered block diagonal (BBD) Jacobian matrix. Our first goal is to compute $A = J_x^{-1} J_u$ and $v = J_x^{-1} g$ in a block-by-block fashion. The simplest case is illustrated in Figure 4a. In this discussion a single prime (') will denote Jacobian and right-hand side elements associated with equation packet "a" and a double prime (") will be used for those associated with "b". The Jacobian matrix associated with this process is shown in Figure 4b. The first step in factoring this matrix is to generate the Jacobian elements associated with unit "a". Nonpivot flags are placed on the equation packet $\Phi$ and on the variable packets ab' and ba' because these packets are shared with a generator that has not yet been processed, and therefore there may be other nonzeros in these columns. The active system is shown in Figure 5a. We then pivot to compute the L/U factors of this matrix to get the structure shown in Figure 5b. The factored block (the non-crosshatched portion in Figure 5b) is stored in one area of mass memory, the residual block (the crosshatched portion) in another. We next repeat this process for "b". The active matrix with the appropriate nonpivot flags is shown in Figure 5c. Pivoting on the

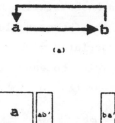

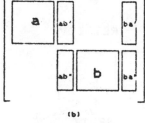

Figure 4. Process containing two interconnected units (a) and its associated Jacobian matrix (b).

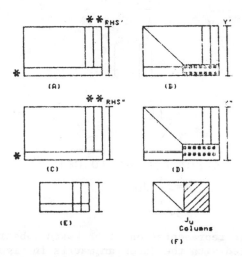

Figure 5. Steps (a) through (f) illustrate the sequence of operations used to factor the Jacobian matrix of Figure 4.

allowable rows and columns leads to the structure represented in Figure 5d. The factored block is sent to mass memory, and the two residual blocks (that is, this one and the one from "a") are combined to give the matrix shown in Figure 5e. There is no need to prevent pivoting in any rows or columns, therefore pivoting leads to the matrix illustrated in Figure 5f. The nonpivoted columns are equivalent to $L^{-1}J_u$ so A is obtained by performing only the backward substitution on these columns. In order to calculate $A = J_x^{-1}J_u$ we move the columns corresponding to $L^{-1}J_u$ to the right hand side of the equality and backsubstitute. Once the backsubstitution has been performed for the residual block, we retrieve the factored block corresponding to 'b'. We identify those columns associated with $J_u$ and move these to the right hand side of the equality. This procedure is shown schematically in Figure 6. The process is separated for 'a'.

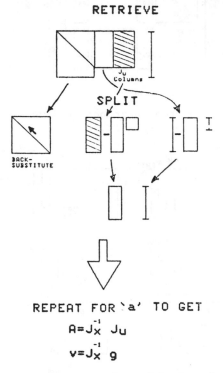

Figure 6. Schematic representation of backward substitution step associated with the Jacobian matrix in Figure 4.

It is a simple matter to show that the following operations can also be performed in a block-by-block manner:

$$K_u + K_x A$$

$$CA \quad ; \quad Cv$$

$$A^T A$$

We have illustrated some of the manipulations that are required; we now examine LASCALA in greater detail keeping in mind the capabilities it must exhibit.

6. LASCALA Commands. As stated in previous sections of this paper, our objective is to develop a problem-oriented language capable of describing certain sparse matrix manipulations. Table 1 contains a list of commands currently proposed for LASCALA. In the present section we describe each of these commands and introduce the computational environment developed to handle large sparse matrices.

TABLE 1

LASCALA Commands

| | |
|---|---|
| Memory Management | Transfer |
| Conversion | Local<br>Global<br>Convert |
| Generating | Set GNGF<br>JGEN<br>KGEN<br>JRGEN<br>KRGEN |
| Solving | Set PNPF<br>Analyze<br>Factor<br>FWD<br>BKWD |
| Other Arithmetic | Add<br>Subt<br>Mult |
| QPP | QPP |

The first command for discussion is the TRANSFER command. To a large extent the efficiency of a sparse matrix scheme is determined by how effectively it uses mass memory and core storage space. On the other hand, one would like to reduce the number of data transfers to mass memory devices because these are very costly operations. On the other hand, one would like to avoid cluttering up core space with data that is not needed for the current matrix calculation (but which may be required later).

Figure 7 illustrates one computational environment which attempts to use core space efficiently while organizing data to reduce the number of transfers to mass memory. The core storage space is divided

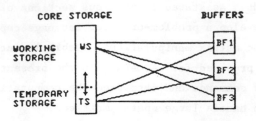

Figure 7. Data structure used with large systems.

into a working store (WS) and a temporary store (TS); the fraction of core allocated to either category can be changed as required by the executive routines. Data which can be sent to mass memory is stored in one of three buffers according to its type. The syntax of the TRANSFER command is as follows:

TRANSFER 'type' 'from' 'to' ('list of names')

where acceptable symbols for "type", and "from" and "to" are given in Table 2.

TABLE 2

Acceptable Parameters for the TRANSFER Command

Data Type Designator

| | |
|---|---|
| VP | Variable Packet |
| FB | Factored Block |
| RB | Residual Block |
| VC | Vector |

Location Designators

| | |
|---|---|
| BF1, BF2, BF3 | Buffers 1, 2 and 3, respectively |
| WS | Working Storage Space |
| TS | Temporary Storage Space |

Within WS is a matrix structure used for core-resident matrix operations. This matrix has a set of LOCAL row and column indices which differ from the set of GLOBAL row and column indices. The commands LOCAL and GLOBAL convert a given set of indices into LOCAL and GLOBAL indices, respectively.

The CONVERT command is used to convert from one sparse matrix data structure to another. At present a matrix may use any of four data structures: full, sparse, rank-one or the data structure peculiar to the user's sparse matrix package. The _sparse_ data structure is a list of the nonzeros in the matrix along with their respective row and column indices. The _rank-one_ data structure contains the k rank-one updates ($w_i$) where the matrix is given by adding the k outer products $w_i w_i^T$ to the identity matrix. The other two data structures are self-explanatory.

The constraint generating commands are SET GNGF JGEN, KGEN, JRGEN and KRGEN. The first command is used to set "generate/no-generate" flags in order to suppress and/or ignore generation of certain rows and/or columns of the Jacobian matrix. The commands JGEN and KGEN are used to generate the partial derivatives associated with the equality and inequality constraints, respectively, while JRGEN and KRGEN are used to generate the residuals of these constraints.

The commands associated with solving a linear system are SET PNPF, ANALYZE, FACTOR, FWD and BACK. The first command is used to set "pivot/no-pivot" flags to indicate any rows or columns whose elements may not be considered for pivot variables. The remaining four commands are used to call the appropriate routines of the user-supplied sparse matrix package described in the Introduction. Other commands which should be compatible with the sparse matrix package are: ADD, SUBT and MULT. These commands are used to add, subtract or multiply two matrices where one or both of the matrices is sparse. The QPP command is used to call the appropriate routines to solve the QPP described in Section 3. Due to the nature of our QPP we do not use a sparse matrix code but rather use a full matrix version of Fletcher's generalized QPP algorithm.

In order to give the reader a more concrete example of how LASCALA might be used, we have included a portion of a sample program. In this program (see Figure 8) each of units A, B, C and D are so large that they must be treated separately in core. The sample program illustrates the sequence of LASCALA commands required to perform the elimination and forward substitution on the linearized equality constraints associated with unit A. A similar sequence of commands is required for the other units and for the backward substitution.

EXAMPLE PROGRAM

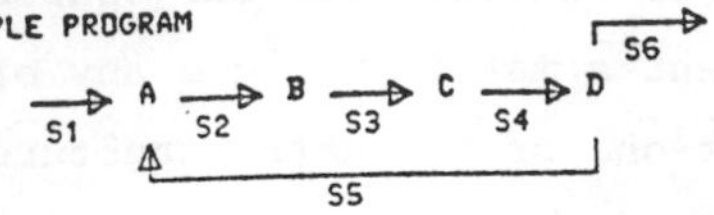

USER INPUT COMPLETE. ALL VARIABLES/EQUATIONS INITIALIZED & SCALED.

C    UNIT A

```
TRANSFER VP BF1 WS (S1,S5,VPA,S2)
TRANSFER VC BF3 WS (LGNGFA,LPNPFA)
SET GNGF (LGNGFA)
SET PNPF (LPNPFA)
JGEN
LOCAL
CONVERT SP SPMP1
ANALYZE
FWD
GLOBAL
TRANSFER FP WS BF2 (VPA)
TRANSFER RB WS TS (RBA)
TRANSFER VP WS TS (S5)
TRANSFER VP WS BF1 (VPA)
TRANSFER VP WS BF1 (S1)
```

Figure 8.  Typical segment from LASCALA program.

**7.  Extensions to LASCALA.**  An obvious extension to the work described thus far, and one which we are currently studying, is to develop a compiler which will automatically generate the necessary
sequence of LASCALA commands associated with the user's description
of his problem.  The compiler should make some attempt to optimize
the order in which the variables are generated/eliminated.  The compiler should also contain guidelines for deciding what order to use
in storing variables for faster retrieval from mass memory.

REFERENCES

Berna, T.J., M.H. Locke and A.W. Westerberg,  A New Approach to Optimization of Chemical Processes,  Paper presented at National
AIChE Meeting, Miami, FL. (November 1978).

Powell, M.J.D.,  A Fast Algorithm for Nonlinear Constrained Optimization Calculations,  Paper presented at the 1977 Dundee Conference on Numerical Analysis.

Westerberg, A.W. and T.J. Berna,  Decomposition of Very Large-Scale
Newton-Raphson Based Flowsheeting Problems,  Computers and Chemical Engineering, 2, 61-63 (1978).

# Practical Comparisons of Codes
# for the Solution
# of Sparse Linear Systems

## Iain S. Duff*

Abstract.   This paper describes some of the practical experience
which the author has had in solving large sparse systems of linear
equations at Harwell and the test bed he has used to examine the
relative performance of different sparse matrix codes.   Most of this
paper is concerned with the direct solution of unsymmetric equations.
We look at the design of a good user interface and the consequent
overheads involved.   We examine various pivoting techniques employed
in sparse Gaussian elimination with particular emphasis on full and
restricted threshold pivoting and the choice of the threshold parameter.
The use of drop tolerances is also discussed.   Comments are made on the
use of interpretative code techniques and hybrid codes.   Possible
gains achieved by the use of preordering techniques are examined.
Some remarks are made on the use of iterative techniques and the
solution of symmetric problems.   The emphasis of this paper is on the
realization of ideas in practical algorithms and they are judged not
merely by their degree of mathematical sophistication but rather by
their effectiveness on practical problems.

1.  Introduction.   In recent years many techniques and heuristics of
varying degrees of sophistication have been suggested and used in codes
for the solution of the set of linear equations

$$A\underline{x} = \underline{b} \qquad\qquad (1.1)$$

where A is large and sparse.   Our purpose here is to discuss some
techniques, mainly those employed in in-core codes for general matrices,
and to critically examine their performance on a wide range of problems
occurring in a variety of different application areas.   In the
following section, we introduce some of the matrices used in our
comparisons and discuss the yardstick by which we will determine the
viability of alternative proposals.

The work for this paper started as a simple exercise in running a new
Harwell code, MA28, on the series of test problems and comparing it with
several other codes in order to evaluate the effectiveness of this new

*AERE Harwell, Didcot, Oxfordshire, England, OX11 0RA.

piece of software for the solution of general sparse unsymmetric sets
of linear equations.    Some of these early runs were described at the
Gatlinburg symposium at Asilomar in December 1977.    It soon became
clear that a simple comparison between codes was fairly meaningless
since different codes have different facilities and are generally
designed to work well on different problem classes.    This present
study looks at the different facilities offered by the various codes
and finds which particular class of problems exhibit characteristics
that these facilities can exploit.    We describe the codes used in our
comparisons in section 2 also.

Even if the internal logistics of the code are well-designed, it is
of little benefit to the average user if the interface is unduly
complicated and the diagnostics poor.    We discuss such considerations
in section 3.

In section 4 we look at three approaches for designing code to solve
(1.1) namely, a compiled code approach, an interpretative code approach,
and looping indexed approaches.    It is this latter one which we
examine in more detail in section 5 where we look at techniques using
partial pivoting, diagonal pivoting and threshold pivoting.    In section
6 we see how our sparse techniques compare with full code methods and,
in particular, examine hybrid techniques where full code is used when
the active matrix becomes sufficiently dense.    A possible technique
for exploiting a commonly occurring structure is that of preordering to
block triangular form.    We examine the viability of such techniques in
section 7.    In section 8, we make some remarks on the solution of
symmetric systems including the use of iterative techniques.

We do not discuss out-of-core techniques or frontal or band schemes.
A discussion of software for these techniques is given by Duff (1979).
Additionally, we have not considered the handling of structurally
singular systems (including rectangular matrices).    Subroutines TRGB
and MA28, described later, cater for such cases.    All of our runs for
these comparisons have been performed on an IBM 370/168 model 3.    We
have not considered the effect of radically different machine
architectures on the performance of sparse matrix codes.    However, some
experience with the Cray 1 suggests that we may have to modify some of
our techniques should the use of such machines become widespread.    For

instance, while a full matrix code, without modification, runs over
20 times faster on a Cray 1 than on an IBM 370/168 the comparable speed
improvement with MA28 is only a factor of 3.

2. Test bed and codes tested. The formation and use of an adequate
test bed is discussed more fully by Duff and Reid (1979). Here we
describe a subset of our test matrices and consider what measurements
we wish to make when performing the tests employed in the present
comparisons. It should be appreciated that the test matrices shown
here and the runs described later in this paper are selected as
representative of results obtained over a much wider range of problems.
Some different examples have been chosen in the case of symmetric
systems and they are described in section 8.

We present the subset of our test matrices in Table 1.1, where, in
addition to giving the order and number of non-zeros, we also give a
short description of the application area and structure of each problem.
Because we believe it to be a qualitatively significant characteristic,
we also give the number of non-zeros after the matrix has been decomposed using Gaussian elimination with the pivots being chosen by
MA28 with the threshold parameter u (see section 5) set to 0.1. The
source of each of these problems is given by Duff and Reid (1979).

Since the majority of codes tested are for the in-core solution of
(1.1), it is natural to use the amount of storage required by each
routine as a yardstick for comparison. Often this will simply be the
number of non-zeros in the decomposed form but sometimes the extra
indexing information is very significant and so must also be included.
One of the user's other main concerns is often that of time. His own
time may depend on his skill in programming as much as the code which
he uses although he can be greatly helped by the creation of a good user
interface. This we believe is of crucial importance and is discussed
in the next section. The run time of his job is also important (it
may, for example, decree that his job must be scheduled as an overnight
run) and this is considered in our comparisons. Often there is a
trade-off between time and storage and some packages allow the user
to choose which is more important. We discuss an example of this in
section 5. Last, but not least, is the question of accuracy. While
we appreciate that the normal user may not be upset by an error in the

| Order | Non-zeros | Non-zeros after decomposition | Comments |
|---|---|---|---|
| 147 | 2441 | 5095 | Symmetric positive definite matrix arising from Ax=λBx problem in theoretical physics. Non-zeros mostly near diagonal. |
| 1176 | 18552 | 21399 | Symmetric. Original was complex but now has real elements randomly generated between 0 and 1. Almost block diagonal structure. Arises from electrical network problem. |
| 199 | 701 | 1478 | Has zeros on diagonal and of no easily described pattern. Arising from stress analysis calculations. |
| 292 | 2208 | 5708 | Pattern is that of normal matrix from least squares adjustment of survey data. |
| 130 | 1282 | 1342 | Has fairly dense small block well connected to large almost diagonal block. Jacobian matrix of a set of ordinary differential equations arising in laser experiments. |
| 363 | 3279 | 4395 | ⎫ |
| 822 | 4726 | 5829 | ⎬ Linear programming bases. |
| 822 | 4790 | 6653 | ⎭ |
| 541 | 4285 | 12537 | ⎫ Nearly symmetric pattern from chemical kinetics problem. |
| 541 | 4285 | 13660 | ⎬ Second matrix is very ill-conditioned. |

Table 1.1    Description of subset of test matrices.

10th significant figure or, in some instances, even in the 4th, he does
want to know that some of his figures are correct and would normally
like to know how many figures to rely on and on which problem  classes
good answers can be expected.    This criterion is also used in our
subsequent comparisons.    Of course, depending on the computing and
accounting system, these factors need to be weighted appropriately and

other measures may be important, so our conclusions are drawn from significant differences in these quantities only.

We now have defined our test matrices and the measurements we wish to make. Our test bed is complete when we describe the tasks we wish our algorithms to perform. Clearly, the user may just want to solve one problem of the form (1.1) which we call a ONE-OFF problem. Some codes take advantage of this by operating on $\underline{b}$ at the same time as performing forward elimination on A and so can save storage. We feel it should be a trivial matter to solve for several right hand sides at the same time. Often, however, the user wishes to solve a sequence of problems of the form (1.1) and so can benefit by retaining information from his first solution. We call the solution of such subsequent equations SOLVE. Finally, particularly in non-linear problems, the user may want to solve (1.1) with a different matrix which has the same sparsity structure as the first. Again information from the first decomposition can be used to facilitate future ones. We call these subsequent decompositions FACTOR. The phase corresponding to the initial decomposition is termed ANALYZE-FACTOR if a numerical factorization is performed at the same time as the pivot selection and ANALYZE if it is not. We would normally expect SOLVE to be faster than FACTOR which, in turn, is usually faster than ANALYZE or ANALYZE-FACTOR.

Finally, in this section, we describe the codes tested giving a brief description of the type of code and the facilities it offers. In these descriptions and in the rest of the paper, a Markowitz ordering is one which chooses as pivot the non-zero whose product of other non-zeros in its row and other non-zeros in its column is minimum. Threshold pivoting requires that the modulus of the pivot is greater than a present multiple times the modulus of the largest element in its row (or column).

GNSOIN.  Generates loop-free Fortran code. Source:  Gustavson (IBM), card deck.

Needs pivot order as input and performs Crout reduction as described by Gustavson et al (1970). Our version generates Fortran source which must then be compiled; other versions generate machine code directly.

TRGB.    Uses interpretative code approach.  Source:  Hutchison
(Cambridge, U.K.), listing.

From the column with the minimum number of non-zeros, the pivot is
that non-zero with minimum row count.  Only accepts pivots larger than
a fixed absolute user-specified tolerance.  Will also handle rectangular
systems.  Has a good user interface.  An earlier version of this code
is discussed in an appendix of a paper by Bending and Hutchison (1973).

MA28            Source:  Harwell Subroutine Library.

Uses Markowitz with threshold pivoting to generate the pivot order,
the structure of the decomposition, and the numerical factors.  Separate
subroutines perform FACTOR and OPERATE.  Principal features are an
extensive user interface, the optional use of block triangularization
routines and the ability to decompose structurally singular problems
including rectangular systems.   Some ideas motivating the choice of data
structures and design of the package are given by Duff and Reid (1977)
and a description of the code and its performance and use by Duff (1977).

NSPIV.            Source:   Sherman (Texas), tape.

Uses partial pivoting to solve a single set of equations (i.e. ONE-
OFF only).  Has a subroutine which rapidly preorders the rows in order
of increasing number of non-zeros.   The main subroutine then chooses
the numerically largest element as pivot from each row in turn.  Described
by Sherman (1978).

SLMATH.                      Source:   IBM.

Uses Markowitz with threshold pivoting to generate pivot order and
has the option of going over to full code when the active matrix is
sufficiently full.  Will also optionally remove zeros caused by numerical
cancellation from the structure.  It performs the factorization in two
stages, namely symbolic factorization (which requires only the pivot
order and need only be done on the first decomposition) and numeric
factorization.   These subroutines are in the IBM Program Product
SLMATH (IBM (1976)).

SSLEST.                      Source:   Thomsen (Lyngby,Denmark),tape.

Has several unique options including restricting pivoting to the
diagonal (or pivoting in natural order) and searching only a limited

number of rows for a good Markowitz count. Uses threshold pivoting and has an option for dropping from the sparsity structure and subsequent decomposition elements smaller than a user-specified value. These subroutines are described by Zlatev et al (1978).

YSMP. Source: Eisenstat (Yale), tape.

Has a variety of options including a ONE-OFF subroutine and subroutines for using the compressed data storage described by Eisenstat et al (1976). Chooses pivots from the diagonal according to a minimum degree algorithm on $A+A^T$. The numerical values are not considered in this ANALYZE phase. The symbolic factorization routine requires only a pivot order and so could get one from an external source. The use of these subroutines is discussed by Eisenstat et al (1977b).

It is worth making a passing comment that a great deal of effort was often required to mount these packages, partly due to difficulties in reading foreign tapes but sometimes because of non-portable features. Additionally, our confidence that the installation was successful was enhanced when test decks and sample output were provided with the code.

For comparison purposes, runs using full code have also been performed, the subroutines used for these runs being MA21 from the Harwell Subroutine Library and a pre-release version of F01BTF from the NAG library. This latter routine performs block column pivoting for efficiency in a paging environment.

3. User interface. As we mentioned in the introduction, we believe that it is of crucial importance that the codes are easy to use, require no specialist knowledge from the user, allow him flexibility in his data interface, and give good diagnostics in the event of failure. Naturally, we cannot expect to obtain such an interface without the penalty of extra overheads on solution time. We illustrate this by looking at the results in Table 3.1 from runs of MA28 on some of our test problems.

We see that, although the overhead is sometimes hidden by the amount of computation subsequently performed, it can be the case that the user must pay a significant penalty for the ease of use or extra

| ORDER<br>NON-ZEROS | 147<br>2441 | 199<br>701 | 130<br>1282 |
|---|---|---|---|
| ANALYZE-FACTOR<br>    Interface<br>    ANALYZE-FACTOR itself | <br>67<br>1393 | <br>26<br>223 | <br>44<br>273 |
| FACTOR<br>    Interface<br>    FACTOR itself | <br>80<br>173 | <br>26<br>23 | <br>30<br>30 |

Table 3.1   Overheads of user interface (IBM 370/168 m.secs.).

facilities allowed to him.  Although we believe that the user's time
is, and will increasingly become, a more vital resource than CPU time
(or number of processors!) we recognize that there are occasions
when the minimization of CPU time is of prime importance.  That is why,
in the design of the MA28 package, we have kept the interface as a
separate driver routine.  If we remove some of the data checking in
the main entry to FACTOR and do not allow input in any order (which
may be different from the order in which elements were input to
ANALYZE-FACTOR), then a FACTOR code with a similar speed to that of
FACTOR itself in Table 3.1 is possible.  We plan to write such an option.

The Waterloo package SPARSPAK (George and Liu (1978)) which handles
matrices of a symmetric structure has an even simpler interface than
MA28 but we have as yet had no experience in using this and so cannot
comment on it in this present series of comparisons.

4.  Compiled and interpretative code approaches. In the compiled
approach, a preprocessing phase generates  code for solving the system
and further systems of the same sparsity structure.   For example,

The Matrix                        With non-zeros stored in positions

```
x     x                          1       6
   x                                 4
x  x  x                          3   5  2
```

of array A

might cause the following code to be produced by the preprocessing
phase (here we have pivoted down the diagonal in order).

```
A(3) = A(3)/A(1)
A(2) = A(2)-A(3)*A(6)
A(5) = A(5)/A(4)
```

In practice, this code is often generated directly as machine code by this preprocessing stage. We use Gustavson's GNSOIN package to represent this approach in our comparisons.

The interpretative code approach is similar to the previous one but, instead of generating explicit code, it generates a sequence of operation codes and addresses so that, for example, the previous example might become

```
1 3 1
2 2 3 6
1 5 4
```

where the operator codes 1 and 2 are used for division and multiplication-subtraction respectively.   Since there are only very few different types of operations used when performing Gaussian elimination, negation can be used to reduce the size of the operator list.  For example, the operator list for the above code could become

```
-3 1 2 6 -5 4 .
```

TRGB from the chemical engineering department at the University of Cambridge uses a similar scheme to this and is used in our test as an example of a code employing an interpretative code approach.

Since there was no preordering algorithm with the GNSOIN package, we used the ordering produced by a precursor to MA28 which was set to choose the pivots on a Markowitz basis with threshold pivoting.   Since our compiler (IBM Fortran H extended) would not handle the code produced by GNSOIN on the larger problems because it was too long or had statements with too many continuation lines we used estimates generated by the subroutine.   These estimates were corroborated by our runs on smaller examples.  To be fair to the designers of GNSOIN, their main version generates machine code and the user is advised that the Fortran version is only suitable for small problems.

The pivoting technique in TRGB gave us some numerical problems which we discuss more fully in the next section so the results quoted for this subroutine are not necessarily for runs which have made numerical sense.

We illustrate runs using these approaches in Table 4.1 where we include some results from a run of MA28 to assist us in drawing the conclusions which follow.

| Order<br>Non-zeros | 147<br>2441 | 199<br>701 | 822<br>4790 | 541<br>4285 |
|---|---|---|---|---|
| Non-zeros in factors<br>    GNSOIN<br>    TRGB<br>    MA28 | <br>5062<br>5073<br>5095 | <br>1416<br>1483<br>1478 | <br>6491<br>7712<br>6653 | <br>14734<br>16771<br>13660 |
| Storage in K-bytes<br>    GNSOIN code<br>    TRGB operator list[+]<br>    MA28 overhead | <br>800*<br>235<br>20 | <br>85.4<br>20.4<br>5.8 | <br>296*<br>99<br>32 | <br>2084*<br>855<br>44 |
| ANALYZE-FACTOR time<br>    GNSOIN<br>(does not include preordering)<br>    TRGB<br>    MA28 | <br>10753<br><br>3148<br>1261 | <br>1552<br><br>466<br>224 | <br>8542<br><br>5951<br>1498 | <br>29101<br><br>19855<br>4043 |
| FACTOR time<br>    GNSOIN<br>    TRGB<br>    MA28 | <br>80*<br>292<br>271 | <br>5.8<br>25<br>51 | <br>19.6*<br>123<br>272 | <br>187*<br>1079<br>578 |
| OPERATE time<br>    GNSOIN<br>    MA28 | <br>7.3*<br>16.6 | <br>2.5<br>8.1 | <br>10.0*<br>35.0 | <br>22*<br>47 |

* estimates (likely to be low)        + assuming INTEGER*2 words

Table 4.1    Compiled and Interpretative Code Approaches.
(Times in milliseconds on an IBM 370/168).

Our conclusions as to the advisability of using such techniques must necessarily be dependent on the problem being solved and the computing environment involved.  Certainly, if we consider in-core solutions to equations (1.1), then the compiled code approach is only valid for quite small problems since the non-zeros of the factors must be held in addition to the code.  At some loss in efficiency it would be possible to code commonly occurring blocks of code as procedures but

we do not know of any software which does this. Although the
resulting loop-free code for FACTOR and OPERATE is much faster than
MA28, the ANALYZE-FACTOR time is significantly higher because of the
time taken to create and output the loop-free code. It should be noted
that, in our terminology, the ANALYZE-FACTOR entry to GNSOIN is really
only a FACTOR entry since the pivot order must be supplied.
Additionally, for our version of the code, the generated code must be
compiled and times for this have not been included in the table.
Dembart and Erisman (1973) have suggested that this approach is only
viable for extremely sparse systems and our results tend to confirm
this.   They suggest a hybrid approach which uses a compiled code
approach until the active matrix becomes fairly dense when an
interpretative code approach is used.

Although the operator list for TRGB is smaller than the compiled
code of GNSOIN, it is still too long to be considered suitable as an
in-core technique unless the problem is very sparse and does not fill-
in much during decomposition.   The TRGB ANALYZE-FACTOR time is more
comparable with MA28, but it does not have an OPERATE entry and its
FACTOR times are very dependent on the density of initial and
decomposed matrices.   Clearly the TRGB FACTOR is much better than the
MA28 FACTOR when the matrix and its factors are very sparse but is
worse when they are not.   There is a slight deficiency in the sparsity
pivoting and a major problem in the numerical pivoting for TRGB;
these we discuss in the next section.

Two benefits of these approaches are that advantage can be taken of
elements in the matrix which are fixed in value (often with value 1.0)
and that it is relatively easy to incorporate variability typing
(Hachtel (1972)).   An implementation benefit lies in the ease with
which the same code generator can be used to create single or double
precision versions of the FACTOR and SOLVE codes.

5. Looping indexed approach and pivoting strategies. In the looping
indexed approach we perform operations on the matrix while analysing
its structure and generate ordering vectors and indexing arrays which
can be used to facilitate the processing of further matrices of the
same structure.   This approach tends to give more portable code than
the approaches in section 4 where machine dependent contrivances are

often employed to achieve maximum efficiency.

In the first part of this section, we examine a variety of pivoting techniques including that of restricting the pivot search to a preset number of rows or columns. We first study them from the viewpoint of preservation of sparsity and then comment on their numerical properties. Later we look at the effect of dropping non-zeros below a preset value from the structure and comment on the compressed storage scheme of Eisenstat et al (1976) which saves storage at the expense of time.

If we can ignore the numerical values of the non-zeros, then we can derive several benefits. If we do not need the reals when determining the pivot order, any workspace needed by this preordering can be equivalenced with storage subsequently used by the reals. Additionally, our pivot selection criterion can be made much simpler since no numerical testing is performed. Symmetric positive definite systems are amenable to this approach where, additionally, the restriction of pivots to the diagonal further simplifies the pivot selection process. We discuss this in more detail in section 8. The YSMP uses essentially the same pivoting technique on unsymmetric systems where the preordering chooses pivots from the diagonal without regard to their numerical values using a minimum degree algorithm (i.e. diagonal element in row with fewest non-zeros is chosen as pivot)on the symmetric system $A+A^T$. We would expect this ordering to do well on systems which are nearly symmetric and to be stable on diagonally dominant systems. Clearly, although the simpler pivot selection scheme should result in a much faster ANALYZE code, the greater fill-in on very unsymmetric systems will have a counterbalancing effect and will be detrimental to the behaviour of the FACTOR and SOLVE phases. Additionally, a factorization based on the YSMP ordering may break down entirely if the matrix has zeros on its diagonal. Because of this problem the runs in Table 5.1 were performed on the test matrices where a zero diagonal entry was replaced by a non-zero whose value equalled the sum of the off-diagonal elements. In Table 5.1, the norm of the solution is about 1.0 and, for example, 2(-12) means that the $\ell_2$-norm of the error in the solution is $2 \times 10^{-12}$.

| Order<br>Non-zeros | 147<br>2441 | 199<br>701 | 822<br>4790 | 541<br>4285 |
|---|---|---|---|---|
| NZ after decomposition<br>    YSMP<br>    MA28 | <br>4510<br>5095 | <br>6255<br>1478 | <br>94058<br>6653 | <br>13213<br>13660 |
| Time (370/168 msecs) for<br>  ANALYZE<br><br>        YSMP<br>        MA28<br>  FACTOR<br><br>        YSMP<br>        MA28<br>  SOLVE<br><br>        YSMP<br>        MA28 | <br><br><br>456<br>1261<br><br><br>108<br>271<br><br><br>11<br>17 | <br><br><br>962<br>224<br><br><br>258<br>51<br><br><br>15<br>8 | <br><br><br>31885<br>1498<br><br><br>16182<br>272<br><br><br>213<br>35 | <br><br><br>1999<br>4043<br><br><br>295<br>578<br><br><br>32<br>47 |
| ERROR<br>        YSMP<br>        MA28 | <br>$2(-12)$<br>$2(-12)$ | <br>$3(-10)$<br>$1(-12)$ | <br>$8(-10)$<br>$1(-11)$ | <br>$8(-6)$<br>$4(-9)$ |

Table 5.1    Diagonal ordering involving no numerical tests.

Although we see that this pivoting technique  does indeed work well on symmetric systems, it can be quite disastrous on very unsymmetric systems both in terms of storage and numerical accuracy.   However, although our results would lead us to reject such a pivoting approach for general unsymmetric systems, it should be pointed out that near symmetric systems frequently do arise in practice and, provided certain safeguards were built into the numeric factorization, such an approach would be useful in these cases.

If we allow pivoting off the diagonal we have a much more varied choice of strategies, many of which were examined by Duff and Reid (1974). The codes we are presently comparing use three approaches.  SLMATH, SSLEST and MA28 use the Markowitz approach, described in section 2, although SSLEST has an interesting variant on this which we examine later in this section.    The TRGB codes choose the pivot as that non-zero in the column of least non-zeros which has fewest non-zeros in its row (this approach, termed $r_i$ in $c_j$, was not considered by Duff and

Reid (1974)). NSPIV first preorders the matrix rows in order of
increasing number of non-zeros and then pivots within each row in turn
to ensure numerical stability. Both these techniques suffer from an
asymmetry which can cause them to perform quite differently on a matrix
and its transpose (for example on the matrix of order 363, the number
of multiplications in the factorization is 6069 when $r_i$ in $c_j$ is applied
to the matrix and is 12229 when the same pivot selection scheme is used
on the transpose). Duff and Reid (1974) have shown that the row preordering technique can produce vast amounts of fill-in. NSPIV does,
however, have the merit of being very short (only 5984 bytes of compiled
code) and simple and when the fill-in is not too great can be very fast.
We examine the merits of these three techniques in Table 5.2.

| Order<br>Non-zeros | 147<br>2441 | 199<br>701 | 822<br>4790 | 541<br>4285 |
|---|---|---|---|---|
| Non-zeros in decomposed<br>form<br><br>Markowitz (MA28)<br>$r_i$ in $c_j$ (TRGB)<br><br>a priori row ordering<br>(NSPIV). (Only non-<br>zeros for backsubstitut-<br>ion phase) | <br><br>5095<br>5073<br><br>4836 | <br><br>1478<br>1483<br><br>1622 | <br><br>6653<br>7712<br><br>8345 | <br><br>13660<br>16771<br><br>51358 |
| Time for one-off<br>(IBM 370/168 msecs)<br>MA28<br>TRGB<br>NSPIV | <br><br>1278<br>3148<br>1606 | <br><br>232<br>466<br>237 | <br><br>1533<br>5951<br>2038 | <br><br>4090<br>19855<br>40295 |

Table 5.2  Comparison of three ordering strategies.

Although, as already mentioned, the possible difference in performance
of the non-Markowitz techniques on the matrix and its transpose
is rather disconcerting, the $r_i$ in $c_j$ criterion generally gives factors
no more than 20% denser than those produced using a Markowitz ordering.
The a priori method of NSPIV saves storage by performing only the
ONE-OFF phase thereby allowing the multipliers to be discarded, but it
still requires a lot of storage because of the inherently bad properties
of the pivot selection technique. To be fair, however, Sherman (1978)

does recommend using other preordering techniques than the one supplied with his package.

The numerical criterion used by the three Markowitz codes (MA28, SLMATH, and SSLEST) is, in all cases, that of threshold pivoting where a non-zero $a_{ik}$, say, is suitable as pivot if and only if

$$|a_{ik}| \geq u.\max_j |a_{ij}| \qquad (5.1)$$

where u is a preset parameter in the range (0,1) which can be chosen by the user so that a balance between pivoting without considering the values of the non-zeros (u=0.) and pivoting where we obtain the numerical stability of partial pivoting (u=1.) can be found. As might be expected these three codes perform similarly with regard to numerical accuracy and MA28 has been chosen to represent this group in Table 5.3. NSPIV effectively has u=1, so that the largest element in its row is chosen as pivot. TRGB has a fixed absolute pivot tolerance accepting any non-zero exceeding this preset value. All the runs in Table 5.3 were performed in double precision on an IBM 370/168. For TRGB we took the best result over several preset values, typically setting this absolute pivot tolerance to 1(-3) times the largest element in the original matrix.

| Order<br>Non-zeros | 147<br>2441 | 199<br>701 | 822<br>4790 | 541<br>4285 |
|---|---|---|---|---|
| Error in solution<br>(Norm of solution is<br>about 1.) | | | | |
| MA28(u=.1)<br>TRGB<br>NSPIV | 2(-12)<br>3(-6)<br>8(-12) | 1(-12)<br>2(-5)<br>7(-13) | 1(-11)<br>6(-5)<br>5(-13) | 3(-9)<br>5(4)<br>2(-12) |

Table 5.3   Numerical properties of three approaches.

As we might expect the partial pivoting strategy of NSPIV is best. However, it is not invariably so because the considerable extra fill-in caused by its pivoting strategy gives a far greater number of numerical operations and hence much more opportunity for rounding errors to affect the calculation. The instability of the TRGB technique is evident (indeed some other values of the absolute pivot tolerance gave worse results). Before we leave these numerical considerations it is interesting to observe how the value of the threshold parameter u in equation (5.1) affects our balance between sparsity and numerical accuracy. Some runs of MA28 with varying u are displayed in Table 5.4.

| Order<br>Non-zeros | 147<br>2441 | | 199<br>701 | | 822<br>4790 | | 541<br>4285 | |
|---|---|---|---|---|---|---|---|---|
| u | Non-zeros in decomposition (NZ) | Error norm | NZ | ERROR | NZ | ERROR | NZ | ERROR |
| 1(-10) | 4881 | 1(2) | 1350 | 4(-9) | 6474 | 1(-8) | 16553 | 5(15) |
| 1(-4) | 5028 | 3(-9) | 1382 | 4(-9) | 6474 | 1(-8) | 16198 | 5(-2) |
| 1(-2) | 5867 | 4(-10) | 1429 | 2(-11) | 6495 | 2(-10) | 15045 | 4(-6) |
| 1(-1) | 5095 | 2(-12) | 1478 | 1(-12) | 6653 | 1(-11) | 13660 | 3(-9) |
| .25 | 6449 | 3(-12) | 1598 | 8(-13) | 6910 | 4(-12) | 14249 | 8(-11) |
| .5 | 6381 | 2(-12) | 1728 | 5(-13) | 7231 | 1(-12) | 14109 | 8(-11) |
| 1.0 | 6772 | 2(-12) | 1915 | 3(-13) | 8716 | 6(-12) | 16767 | 2(-10) |

Table 5.4   Varying the parameter in threshold pivoting.

The stability of the method is essentially monotonic with respect to u although the higher number of non-zeros in the factors (and more arithmetic operations) when u=1. can cause the error to increase in that case.   It appears that an optimal level of fill-in is reached quite early when decreasing u from 1.0.   Indeed at very low values of u, the fill-in can start increasing because the growth in size of elements tends to be proportional to $u^2$ rather than u itself which may make non-zeros best suited on sparsity grounds unavailable as pivots. We have compromised by recommending a value of u=0.1 in Duff (1977). The recommended value in SSLEST is $1/16$.

Zlatev et al (1978) have recommended restricting the pivot choice to a few rows (typically 3 or 4) with the smallest number of non-zeros. We have not adopted this approach in MA28 because we find we do not normally search many rows and columns before finding our best Markowitz pivot.   The figure for the 199 case in Table 5.5 is fairly extreme, those in the other two cases being more typical.

We show in Table 5.6 the results from running SSLEST with the Markowitz search restricted to differing numbers of rows.   We see that sometimes our results support the claims of Zlatev et al (1978) but there is clearly no optimal number of rows to search with, in general,

| Order<br>Non-zeros | 147<br>2441 | 199<br>701 | 292<br>2208 |
|---|---|---|---|
| Average no. of rows<br>and columns searched | 5.0 | 14.5 | 3.1 |

Table 5.5    Amount of searching in MA28 for best Markowitz pivot.

| Order, N<br>Non-zeros | | 147<br>2441 | 199<br>701 | 292<br>2208 | 541<br>4285 |
|---|---|---|---|---|---|
| Non-zeros in<br>decomposition | | | | | |
| Restricted<br>to this<br>number<br>of sparsest<br>rows | 1 | 4517 | 1407 | 5750 | 13727 |
| | 2 | 4509 | 1380 | 5112 | 12978 |
| | 3 | 4420 | 1342 | 5153 | 13252 |
| | 4 | 4480 | 1417 | 5228 | 13211 |
| | 5 | 4465 | 1388 | 4956 | 13204 |
| | N | 4516 | 1244 | 5305 | 12540 |

Table 5.6    Restricting pivot selection to sparsest rows.

a full Markowitz search yielding the sparsest factors. The times for all the restricted runs in Table 5.6 were comparable to the MA28 ANALYZE-FACTOR phase but when SSLEST was asked to obtain the best Markowitz pivot these times were increased fourfold or more. We feel that this is more a reflection on the design of the code (and we note that SSLEST does not recommend unrestricted Markowitz) rather than a defect in the pivoting strategy and argue that not only is a strategy which chooses the best Markowitz pivot aesthetically satisfying but it is also computationally feasible.

Zlatev and Nielsen (1977) have recommended ignoring small fill-ins when performing the original decomposition and then using iterative refinement to improve the solution. Although we have not performed iterative refinement, we have run SSLEST with various drop tolerances and present the results in Table 5.7. We have not done any runs with the tolerance greater than $10^{-2}$ since Zlatev and Nielsen reported problems in the convergence of iterative refinement if this were done and we already see a marked increase in error at this stage. Again the $\ell_2$-norm of our solution is about 1.0.

| Order<br>Non-zeros | 1176<br>18552 | | 199<br>701 | | 822<br>4726 | | 541<br>4285 | |
|---|---|---|---|---|---|---|---|---|
| Drop Tolerance | Non-zeros in factors (NZ) | ERROR | NZ | ERROR | NZ | ERROR | NZ | ERROR |
| 1(-8) | 18757 | 9(-13) | 1337 | 7(-7) | 6241 | 1(-11) | 11965 | 2(-4) |
| 1(-6) | 18757 | 1(-5) | 1338 | 3(-4) | 6239 | 1(-4) | 12575 | 4(-3) |
| 1(-4) | 18746 | 1(-4) | 1335 | 2(-1) | 6094 | 7(-2) | 10054 | 8(-1) |
| 1(-2) | 18730 | 3(-1) | 1114 | 4(2) | 5768 | 2(1) | 7202 | 1(2) |

Table 5.7   Effect of varying the drop tolerance.

Our main comment on the results of Table 5.7 is that although the drop tolerance appears to have a strong effect on the accuracy of the decomposition, it does not generally have a very significant effect on the number of non-zeros in the decomposition or on the time for ANALYZE-FACTOR (not shown in table).   However, there are problems for which large savings in storage can be made without such bad numerical deterioration (for example, Zlatev and Nielsen (1977)).  We use a partial factorization technique for this type of symmetric system in section 8.

In many instances, when working with sparse systems, there is a fine balance between time and storage and it is often possible to trade off one for the other.   The example we choose here is that of the compressed storage scheme as used by the YSMP and described by Eisenstat et al (1976).  We show, in Table 5.8, runs of the YSMP with and without this compressed storage scheme.

The trade-off is evident.  Clearly it is up to the user to decide whether he wants to incur a time penalty in order to save storage. Sherman (private communication) reports that the compressed scheme does comparably far better in the symmetric version of their codes when run on fairly regular problems from finite difference or finite element discretizations.

| Order<br>Non-zeros | 147<br>2441 | 199*<br>701 | 363*<br>3279 | 541<br>4285 |
|---|---|---|---|---|
| **Total words of storage** | | | | |
| NDRV | 9462 | 13108 | 49678 | 28050 |
| CDRV | 6962 | 9740 | 34642 | 24153 |
| **Time:** | | | | |
| ANALYZE | | | | |
| NDRV | 456 | 962 | 5712 | 1999 |
| CDRV | 602 | 1141 | 6378 | 2267 |
| FACTOR | | | | |
| NDRV | 108 | 258 | 1880 | 295 |
| CDRV | 153 | 350 | 2462 | 506 |
| SOLVE | | | | |
| NDRV | 11 | 15 | 54 | 32 |
| CDRV | 13 | 19 | 67 | 41 |

*
Diagonals made non-zero.

Table 5.8    The use of a compressed storage scheme. (NDRV uses the uncompressed scheme while CDRV uses the compressed storage scheme.)

6.  The use of full code. Sparse code is very much more complicated than full code for performing Gaussian elimination.    One illustration of this is seen in the 8.9 Kbytes of code for the Harwell full matrix code MA21 as opposed to the complete MA28 package which compiles (without overlay) to 38 Kbytes.  Certainly full code is better for completely full systems and so it is interesting to compare its performance with that of a sparse code on some of our test matrices.

On all our test matrices, a few of which are shown in Table 6.1, the full codes are worse than MA28 for the ANALYZE-FACTOR phase both in terms of work and storage.  Additionally, even when ANALYZE-FACTOR is competitive, the sparse FACTOR and SOLVE take so much advantage of the sparsity present in the factors that the full code is totally un-competitive there.   Of course, if the matrix were full then there is no advantage in using sparse techniques.   However, this is exactly what happens to the active matrix during sparse Gaussian elimination so there is normally a point during the decomposition where it is more efficient to switch to full code.  SLMATH allows us to go over to full code during ANALYZE-FACTOR when the active matrix reaches a preset

| Order:  N | 147 | 199 | 130 | 541 |
|---|---|---|---|---|
| Non-zeros | 2441 | 701 | 1282 | 4285 |
| $N^2$ | 21609 | 39601 | 16900 | 292681 |
| Non-zeros in factors | 5095 | 1478 | 1342 | 12537 |
| <u>TIME</u> (370/168 m.secs.) | | | | |
| ANALYZE-FACTOR | | | | |
| MA28 | 1261 | 224 | 285 | 3032 |
| MA21 | 2440 | 6547 | 1690 | 160740 |
| F01BTF | 2050 | 5033 | 1513 | 100473 |
| FACTOR | | | | |
| MA28 | 271 | 51 | 70 | 503 |
| MA21 | 2440 | 6547 | 1690 | 160740 |
| F01BTF | 2050 | 5033 | 1513 | 100473 |
| SOLVE | | | | |
| MA28 | 17 | 8 | 6 | 44 |
| MA21 | 50 | 100 | 37 | 850 |
| F04AYF | 43 | 77 | 40 | 930 |

Table 6.1    A comparison of a full code with MA28.

density (and is of a certain preset minimum size) and we have used this package to produce the results in Table 6.2.    In this table, a density of 101% means that we continue using sparse matrix techniques to the end of the elimination even if the active matrix is full.

The FACTOR and SOLVE subroutines only use the pivot sequence generated in ANALYZE-FACTOR and so themselves do no full matrix processing.  If they did so then the balance would be altered, probably increasing the benefits of using full code at the higher densities but incurring more costly overheads when the active block is quite large and sparse.    These runs indicate that even for quite low densities it is worthwhile switching to full matrix code.    The gains can be quite significant although they will to some extent reflect the efficiency of the sparse code. We include the MA28 ANALYZE-FACTOR times to substantiate this comment.

| Order<br>Non-zeros | | 147<br>1441 | 199<br>701 | 822<br>4790 | 541<br>4285 |
|---|---|---|---|---|---|
| ANALYZE-FACTOR TIME | | | | | |
| | MA28 | 1261 | 224 | 1498 | 3032 |
| Density of<br>active matrix<br>when switch<br>made | 101 | 2990 | 441 | 2976 | 15428 |
| | 100 | 2611 | 413 | 2926 | 15155 |
| | 80 | 2342 | 400 | 2932 | 14182 |
| | 60 | 2171 | 388 | 2944 | 13357 |
| | 40 | 2089 | 379 | 2947 | 12342 |
| | 20 | 2862 | 376 | 2993 | 16049 |
| FACTOR TIME | | | | | |
| | 101 | 115 | 19 | 77 | 375 |
| | 100 | 112 | 18 | 75 | 366 |
| | 80 | 121 | 19 | 76 | 381 |
| | 60 | 127 | 20 | 77 | 381 |
| | 40 | 154 | 21 | 83 | 445 |
| | 20 | 372 | 25 | 99 | 1453 |
| SOLVE TIME | | | | | |
| | 101 | 16 | 7 | 30 | 52 |
| | 100 | 16 | 7 | 30 | 51 |
| | 80 | 17 | 7 | 30 | 52 |
| | 60 | 17 | 7 | 31 | 52 |
| | 40 | 18 | 7 | 32 | 54 |
| | 20 | 26 | 8 | 34 | 81 |

Table 6.2   Use of full matrix code by SLMATH at end of sparse
elimination.   Times in m.secs. on an IBM 370/168.

7.  Preordering to block triangular form.   If it is possible to
order the matrix to the block triangular form

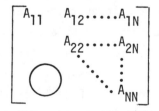

where the submatrices $A_{ii}$ (i=1,...,N) are square then a solution to the
whole system can be found by solving subsystems with coefficient
matrices $A_{ii}$ (i=N,...,1) consecutively.   Gaussian elimination is
performed only within the diagonal blocks and the off-diagonal blocks

are used only for forward substitution. The cost of such a preordering is quite low. Some examples are shown in Table 7.1.

| Order of matrix<br>Number of non-zeros | 147<br>2441 | 199<br>701 | 292<br>2208 | 822<br>4726 |
|---|---|---|---|---|
| BLOCK<br>ANALYZE-FACTOR<br>FACTOR<br>SOLVE | 80<br>1380<br>270<br>20 | 30<br>210<br>50<br>10 | 100<br>960<br>210<br>20 | 260<br>470<br>210<br>40 |

Table 7.1    Block triangularization times.

In fact only the second and fourth of the cases have a non-trivial block structure. Thus in cases 1 and 3 nothing is gained by calling BLOCK, but it can be seen that the computing overhead is quite slight. In the second case the overhead happens to be exactly recovered in faster execution of ANALYZE-FACTOR. The last case has a very substantial block structure and without its recognition the ANALYZE-FACTOR time rises to 1500 m.secs.

Block triangularization is an optional feature in the Harwell code MA28 with which the above runs were performed.

8. Symmetric systems. Several other papers at this conference deal specifically with structurally symmetric systems so we will not make many detailed comments in this contribution. A slightly fuller review of techniques applicable to symmetric matrices is given by Duff (1979).

The test matrices used in the comparisons of this section have been obtained from three sources. One set was obtained from the matrices used previously by reflecting the lower triangular pattern of non-zeros in the diagonal. These were augmented by a few randomly generated symmetric matrices. Another set was obtained from Alan George (Waterloo) and consists of matrices arising from various finite element triangulations of an L-shaped region. Except for the matrix of order 147, the non-zero values for both these sets were pseudo-random numbers in the range (-1,1). Our final set arises from finite difference discretizations of Laplace's equation in a square or cube using the 5 and 7 point formulae respectively.

A recent exciting development in decomposing symmetric matrices has come from the use of techniques, originally employed in systems arising from finite element situations, which use a generalized element model and view the decomposition as a concurrent sequence of amalgamations of elements into superelements and elimination of variables contained entirely within such superelements (for example, Eisenstat et al (1979) and George and Liu (1979)). A great advantage of such an approach is that it works with the initial sparsity pattern and indexing vectors of length determined by the order of the system. This results (particularly when fill-in is high) in an ANALYZE time which is much less than the subsequent FACTOR time (quite the opposite from our experience on un-symmetric systems) and additionally allows the ANALYZE step to be done in a predetermined amount of storage.

Naturally, we would expect the generalized element approach to do particularly well on matrices arising from finite element discretizations and so we have classified the matrices in Table 8.1 according to whether they arose from finite element problems or not. The minimum degree algorithm is a symmetric variant of the YSMP described in section 2 (Eisenstat et al (1977a)) and Harwell subroutine MA31A (Munksgaard (1979)) for the element and non-element problems respectively, and the generalized elements algorithm was written by John Reid (Harwell) and uses a minimum degree criterion to order the assemblies. The matrices of order 1270 and 3466 come from the set of problems from the L-shaped region while the rest are symmetrized versions of our unsymmetric set.

| | Finite Element Problems | | Non-Element Problems | | | |
|---|---|---|---|---|---|---|
| ORDER OF MATRIX | 1270 | 3466 | 1176 | 199 | 130 | 363 |
| NON-ZEROS IN LOWER TRIANGLE (INCLUDING DIAGONAL) | 4969 | 13681 | 9864 | 536 | 713 | 1863 |
| Minimum degree | 1551 | 6048 | 1475 | 86 | 83 | 496 |
| Generalized elements | 730 | 2047 | 240 | 90 | 173 | 667 |

Table 8.1    ANALYZE time (370/168 m.secs) for generalized element method and minimum degree algorithm.

The vast superiority of a generalized element approach when used in the ANALYZE phase of finite element problems is quite evident. However, it is most interesting to see that this approach is often very competitive when used on systems which are not from finite element problems. The great gain in the case of order 1176 is due to the fact that its structure is very similar to that of a matrix arising from a finite element discretization. The approach is still in its infancy and it will be interesting to see whether it becomes the accepted method of performing the ANALYZE phase for general symmetric systems.

In section 6, we discussed the gains that could be achieved by switching to a full matrix code towards the end of the elimination. However, we can realize much greater gains when switching to full matrix code during the decomposition of positive definite systems since, when the switch is made, our ANALYZE phase is essentially finished! We have only to complete our permutation array and we are done. A code by Munksgaard (1979) incorporates this idea and we compare his code with the symmetric YSMP in Table 8.2.

| Order | | 147 | 199 | 363 | 400 |
|---|---|---|---|---|---|
| Non-zeros in lower triangle (including diagonal) | | 1441 | 536 | 1962 | 1199 |
| Non-zeros in factors | | | | | |
| Density of active matrix when switch performed | 100 | 2348 | 1074 | 4725 | 5768 |
| | 90 | 2403 | 1084 | 4854 | 5831 |
| | 80 | 2446 | 1102 | 5174 | 5914 |
| Order of active matrix when switch performed | | | | | |
| | 100 | 29 | 19 | 51 | 83 |
| | 90 | 35 | 21 | 64 | 86 |
| | 80 | 39 | 24 | 75 | 89 |
| Time for one-off (370/168 m.secs) | | | | | |
| | YSMP | 320 | 142 | 1044 | 1041 |
| | 100 | 477 | 140 | 1155 | 987 |
| | 90 | 430 | 127 | 983 | 805 |
| | 80 | 399 | 123 | 830 | 720 |

Table 8.2   Recognition of dense active matrix.

Significant gains can be obtained by performing this switch. It should be pointed out in passing that the generalized element approach lends itself to allowing the use of full matrix codes in the inner loops.

Finally, some interesting developments are afoot which are blurring the distinction between direct and iterative methods. We refer to the

use of approximate factorizations of A which are then used as a preconditioning technique for an iterative scheme. We saw one instance
of this in section 5 where a partial factorization (admittedly a fairly
accurate one) was used with iterative refinement in the solution of
unsymmetric equations. Here we consider preconditioning for the
conjugate gradient algorithm. An example of such a technique is the
ICCG method (Meijerink and van der Vorst (1977)). Munksgaard (1979)
has developed an algorithm to calculate as accurate factors of the
inverse as can be obtained in a predetermined space and uses this as
his preconditioning matrix. We give some of his results in Table 8.3,
where we observe that a direct method and the ICCG algorithm are just
the two extremes of his technique. The matrix of order 700 was
randomly generated while the others were symmetrized versions of our
unsymmetric set.

| Order<br>Off-diagonal non-zeros<br>(including diagonal) | 363<br>1962 | 822<br>2803 | 700<br>2099 |
|---|---|---|---|
| Non-zeros in factors<br>    ICCG<br>    MID-WAY<br>    DIRECT | 1962<br>2421<br>5174 | 2803<br>3506<br>12973 | 2099<br>3425<br>16084 |
| TIME (370/168 m.secs)<br>  ANALYZE<br>    ICCG<br>    MID-WAY<br>    DIRECT | <br><br>488<br>578<br>804 | <br><br>378<br>501<br>1991 | <br><br>248<br>531<br>2375 |
| SOLVE    ICCG*<br>    MIDWAY*<br>    DIRECT | 494(18)<br>244(7)<br>26 | 943(43)<br>406(7)<br>65 | 536(14)<br>318(6)<br>86 |

*The numbers in brackets in these rows give the number of
conjugate gradient iterations required to reduce residual to $10^{-8}$.

Table 8.3  Preconditioned conjugate gradient algorithm.

Since techniques akin to the ICCG have been developed primarily for
matrices arising from the solution of partial differential equations, we
have run Munksgaard's code, Harwell subroutine MA31, on matrices from
this area. The first two are from the 5-point formula used to
discretize Laplace's equation on grids of order 16x16, 32x32,

respectively while the last two are from the 7-point discretization
on grids of size 8x8x8 and 10x10x10 respectively.

| Order<br>Number of non-zeros<br>(including diagonal) | 256<br>736 | 1024<br>3008 | 512<br>1856 | 1000<br>3700 |
|---|---|---|---|---|
| Number of non-zeros in factors | | | | |
|     ICCG | 736 | 3008 | 1856 | 3700 |
|     MID-WAY | 1356 | 6113 | 3691 | 7590 |
|     DIRECT | 2227 | 12559 | 13124 | 35086 |
| TIME (m.secs.on an IBM 370/168) | | | | |
| ANALYZE | | | | |
|     ICCG | 81 | 333 | 237 | 463 |
|     MID-WAY | 180 | 893 | 569 | 1152 |
|     DIRECT | 285 | 2387 | 2779 | 11505 |
| SOLVE | | | | |
|     ICCG | 169(15) | 1029(19) | 353(11) | 752(12) |
|     MID-WAY | 84(5) | 637(6) | 243(5) | 571(6) |
|     DIRECT | 12 | 67 | 61 | 153 |

Table 8.4    Direct and iterative methods on finite difference
matrices.    Numbers in brackets as for Table 8.3.

These results confirm the normally accepted belief that where
applicable iterative methods should be used for 3-dimensional problems
while the two approaches are quite comparable in the two-dimensional
case with direct methods superior on smaller problems.    The above
results also indicate that a direct method will be better if we have to
solve many systems with the same coefficient matrix.

However, the distinction between direct and iterative methods is no
longer so clear cut as it once was and the semi-direct/semi-iterative
methods could become the best techniques for such problems.

REFERENCES

[1] M.J. BENDING and H.P. HUTCHISON. The calculation of steady state incompressible flow in large networks of pipes, Chem. Eng. Sci. 28, pp.1857-1864 (1973).

[2] B. DEMBART and A.M. ERISMAN. Hybrid sparse-matrix methods, IEEE Trans. Circuit Theory CT 20 pp.641-649 (1973).

[3] I.S. DUFF. MA28 - a set of Fortran subroutines for sparse unsymmetric linear equations, AERE Report R.8730, HMSO, London (1977).

[4] I.S. DUFF. Some current approaches to the solution of large sparse systems of linear equations, Harwell Report CSS 65, (1979). Presented at Int'l Congress on Numer. Meth. Engng. Paris, Nov. 1978.

[5] I.S. DUFF and J.K. REID. A comparison of sparsity orderings for obtaining a pivotal sequence in Gaussian elimination, J. Inst. Math. Appl. 14, pp.281-291 (1974).

[6] I.S. DUFF and J.K. REID. Some design features of a sparse matrix code, Harwell Report CSS 48 (1977). To appear in TOMS.

[7] I.S. DUFF and J.K. REID. Performance evaluation of codes for sparse matrix problems, Harwell Report CSS 66 (1979). To appear in Proc. of Conf. on Performance Evaluation of Numerical Software, Baden, Austria, Dec. 1978. Edited by L. Fosdick, North Holland.

[8] S.C. EISENSTAT , M.H. SCHULTZ and A.H. SHERMAN. Considerations in the design of software for sparse Gaussian elimination, in Sparse Matrix Computations, edited by Bunch and Rose, Academic Press (1976).

[9] S.C. EISENSTAT , M.C. GURSKY , M.H. SCHULTZ and A.H. SHERMAN. Yale sparse matrix package. I. The Symmetric codes. Research Report H 112, Dept. of Computer Science, Yale University (1977a).

[10] S.C. EISENSTAT , M.C. GURSKY , M.H. SCHULTZ and A.H. SHERMAN. Yale sparse matrix package. II. The Non-symmetric codes. Research Report H 114, Dept. of Computer Science, Yale University (1977b).

[11] S.C. EISENSTAT , M.H. SCHULTZ and A.H. SHERMAN, Software for sparse Gaussian elimination with limited core storage, These proceedings (1979).

[12] J.A. GEORGE and J.W.H. LIU. User Guide for SPARSPAK: Waterloo Sparse Linear Equations Package. Comput. Sci. Report CS-78-30 Waterloo, Canada (1978).

[13] J.A. GEORGE and J.W.H. LIU. A quotient graph model for symmetric factorization. These proceedings (1979).

[14] F.G. GUSTAVSON , W.M. LINIGER and R.A. WILLOUGHBY, Symbolic generation of an optimal Crout algorithm for sparse systems of linear equations, J. Assoc. Comp. Mach. 17, pp.87-109 (1970).

[15] G.D. HACHTEL, Vector and matrix variability type in sparse matrix algorithms, in Sparse matrices and their applications, edited by Rose and Willoughby, Plenum Press, pp.53-64 (1972).

[16] IBM, IBM System/360 and System/370, IBM 1130 and IBM 1800 Subroutine Library - Mathematics. User's Guide. Program Product 5736-XM7. IBM Catalogue SH12-5300-1 (1976).

[17] J.A. MEIJERINK and H.A. VAN DER VORST, An iterative solution method for linear systems of which the coefficient matrix is a symmetric M-matrix. Math. Comp. 31, pp.148-162 (1977).

[18] N. MUNKSGAARD, Solving sparse symmetric sets of linear equations by preconditioned conjugate gradients, Harwell Report CSS 67 (1979).

[19] A.H. SHERMAN, Algorithms 533. NSPIV, A Fortran subroutine for sparse Gaussian elimination with partial pivoting, TOMS 4, pp.391-398 (1978).

[20] Z. ZLATEV , V.A. BARKER and P.G. THOMSEN, SSLEST: A Fortran IV subroutine for solving sparse systems of linear equations. User's Guide. Tech. Rep. 78-01, Numersk Inst., Lyngby, Denmark (1978).

[21] Z. ZLATEV and H.B. NIELSEN, Preservation of sparsity in connection with iterative refinement, Tech. Report NI-77-12. Numersk. Inst. Lyngby, Denmark (1977).

# Software
# for Sparse Gaussian Elimination
# with Limited Core Storage*

## S. C. Eisenstat†, M. H. Schultz†, and A. H. Sherman‡

Abstract. A variant of Gaussian elimination is presented
for solving sparse symmetric systems of linear equations on
computers with limited core storage, without the use of
auxiliary storage such as disk or tape. The method is based
on the somewhat unusual idea of recomputing rather than
saving most nonzero entries in the reduced triangular system, thus trading an increase in work for a decrease in
storage. For a nine-point problem with the nested dissection ordering on an n x n grid, fewer than $\frac{7}{2}n^2$ nonzeroes
must be saved versus $\sim\frac{93}{12}n^2\log_2 n$ for sparse elimination,
while the work required at most doubles. The use of auxiliary storage in sparse elimination is also discussed.

1. Introduction. Consider the system of linear equations

(S)    A x = b

where the coefficient matrix A is a sparse N x N symmetric

positive definite matrix such as arise in finite-difference

and finite-element approximations to elliptic boundary-

value problems. Direct methods for solving (S) are general-

ly variations of (symmetric) Gaussian elimination:  We use

*This research was supported in part by ONR Grant N00014-
76-C-0277 and AFOSR Grant F49620-77-C-0037.

†Department of Computer Science, Yale University, New
Haven, Connecticut 06520.

††Department of Computer Science, The University of Texas
at Austin, Austin, Texas 78712.

135

the $k^{th}$ equation to eliminate the $k^{th}$ variable from the remaining N-k equations for k = 1,...,N, and then solve the resulting triangular system (the reduced system) for x. Unfortunately, as the elimination proceeds, some coefficients that were zero in the original system of equations become non-zero (fill-in), increasing the work and storage required.

As an example, consider the following model problem which arises from the familiar nine-point finite-difference discretization of the Poisson equation on the unit square with homogeneous Dirichlet boundary conditions. Given a uniform n x n grid in the plane (see Figure 1), we associate a variable $u_{ij}$ with each mesh-point (i,j) and form the system of linear equations

$$8u_{ij} - u_{i-1,j-1} - u_{i-1,j} - u_{i-1,j+1} - u_{i,j-1}$$
$$- u_{i+1,j+1} - u_{i+1,j} - u_{i+1,j-1} - u_{i,j+1} = f_{ij} \quad 1 \le i,j \le n$$

where

$$u_{ij} = 0 \quad i = 0, \text{n+1} \quad \text{or} \quad j = 0, \text{n+1}$$

There are $N = n^2$ unknowns $u_{ij}$ and, taking account of symmetry, $\sim 6n^2$ nonzeroes in A and b. But with the nested dissection ordering of the variables [7] (which is optimal to within a constant factor [8]), the number of nonzeroes in the reduced system is $\sim \frac{93}{12}n^2\log_2 n$.

Our model problem illustrates the behavior one often encounters in using Gaussian elimination to solve large sparse systems: The storage required can easily exceed

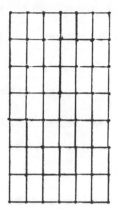

Figure 1

the core storage available for even moderately large N, even though the problem and solution (i.e., A, b, and x) <u>can</u> be represented in core. Thus, although we could store the nonzero coefficients, right-hand side, and solution for our model problem in $\sim 7n^2$ locations, the reduced system would require an additional $\sim \frac{93}{12}n^2 \log_2 n$ locations.

In this paper we discuss variants of sparse elimination which can solve (S) with minimal core storage. The methods are based on the following assumptions:

(1)  The nonzero entries of A and b are inexpensive to generate on demand (e.g., they can be stored in core or computed with little effort).

(2)  There is enough core storage for the solution vector x, plus a small amount of working storage, but not enough to store the entire reduced system.

We count only the working storage required to solve (S) in stating the storage requirements of such methods. All other

storage (i.e., storage for A, b, and x) is associated
with the linear system rather than the method of solution
and is ignored.

The standard solution is not to store the entire reduced
system in memory, but to use auxiliary storage (e.g., disk
or tape) as well.[1] An alternate approach is not to store
the entire reduced system at all (cf. [4]). Instead, we
throw away most nonzero entries and recompute them as necessary during back-solution. The result is an algorithm
which can solve our model problem with approximately twice
as much work as sparse elimination, but which requires
that fewer than $\frac{7}{2}n^2$ nonzero entries be stored, versus
$\sim\frac{93}{12}n^2 \log_2 n$ for the reduced system.

In Section 2, we review the nested dissection ordering
[7] and the "Disaster Strikes" algorithm for solving the
model problem [3]. We introduce an element model of elimination in Section 3 in order to generalize the algorithm
to non-model problems, and show how to implement such a
scheme in Section 4. In Section 5, we present some experimental results, and in Section 6 the applications of
these ideas to auxiliary storage sparse elimination.

2. The Nested Dissection Ordering. The work and storage required to solve a large sparse system of linear
equations clearly depend upon the zero-nonzero structure
of the coefficient matrix. But since this matrix is posi-

---

[1] Virtual memory systems appear to have large amounts of
memory but in reality constitute a hidden use of auxiliary
storage.

tive definite and symmetric, we could equally well solve the permuted system

$$PAP^t \; y = P \; b, \quad P \; x = y$$

given any permutation matrix P [11]. The permuted system corresponds to reordering the variables and equations of the original system, and the net result can often be a significant reduction in the work and storage required to form the reduced system. For the case of the nine-point operator, George [7] has discovered a nearly optimal ordering known as the nested dissection ordering.

The easiest way to describe the nested dissection ordering is in reverse order; that is, we shall describe the last group of variables to be eliminated, then the second last, and so forth. The exact order in which variables in each group are eliminated does make some difference to the total work and storage required, but we shall ignore this difference. Although George described his ordering in terms of independent sets, we shall use the recursive description given by Rose and Whitten [10].

The basic step consists of numbering the 2n+1 variables on a central dividing cross (see Figure 2). These unknowns are the last to be eliminated. When we delete them, the grid splits into four $\sim\frac{n}{2}$ x $\sim\frac{n}{2}$ subgrids. Each of these subgrids is structurally similar to the original grid so that we can number it in the same fashion.

Theorem. (George [7]) For the nine-point model problem with the nested dissection ordering on an n x n grid, sparse elimination requires

$$0_{ND}(n) = 10n^3 + 0(n^2 \log_2 n)$$

multiply-adds, and the reduced system has

$$S_{ND}(n) = \frac{93}{12} n^2 \log_2 n + 0(n^2)$$

nonzero entries.

```
o o o x o o o            o x o   o x o

o o o x o o o            x x x   x x x

o o o x o o o            o x o   o x o

x x x x x x x   ----->

o o o x o o o            o x o   o x o

o o o x o o o            x x x   x x x

o o o x o o o            o x o   o x o
```

Figure 2

When the last variable has been eliminated, we have generated all the nonzero entries in the reduced system of equations that remains, and have merely to solve this system for the vector of unknowns x. Yet suppose disaster strikes and only those entries in the last 2n+1 rows are saved. Then we could still solve for the last 2n+1 variables, i.e., the values of the unknowns on the central dividing cross. But, given the values of these variables, our n x n problem splits into four smaller $\sim\frac{n}{2}$ x $\sim\frac{n}{2}$ problems of the same form. These subproblems can now be solved in the same fashion.

"Disaster Strikes" Algorithm [3]:

(1) Solve for the unknowns on a central dividing cross.

(2)   The problem splits; solve each of the subproblems in the same fashion.

Of course, throwing away nonzero entries means that we will have to do some additional work to recompute them. But how much more?  The cost of solving an n x n problem is just $O_{ND}(n)$ plus the cost of solving four $\sim\frac{n}{2}$ x $\sim\frac{n}{2}$ problems.  Letting $O_{MS}(n)$ denote this cost, we have

$$O_{MS}(n) = O_{ND}(n) + 4\ O_{MS}(\sim\tfrac{n}{2})$$
$$= 10n^3 + 4\ O_{MS}(\sim\tfrac{n}{2}) + O(n^2\log_2 n)\ .$$

Thus,

$$O_{MS}(n) = 20n^3 + O(n^2\log_2 n)\ ,$$

and we are doing approximately twice as much work.  But, as we shall see, the saving in storage is more significant.

3.  An Element Model of Elimination.  How much storage is required to perform sparse elimination on our model problem with the nested dissection ordering if we only need to save the nonzero entries in the last 2n+1 rows of the reduced system?  How do we generalize the "Disaster Strikes" algorithm to non-model problems?  How do we implement such an algorithm?  In this section, we introduce an element model of elimination which will help to resolve these questions.

The element model emulates Gaussian elimination by a sequence of transformations on a collection $E$ of sets of variables called elements.  Initially,

$$E^{(0)} = \{\{x_i, x_j\} \mid a_{ij} \neq 0, \ i \leq j\} \ .$$

Corresponding to using the $k^{th}$ equation to eliminate the $k^{th}$ variable from the remaining N-k equations, we transform $E^{(k-1)}$ to $E^{(k)}$ by

(1)  merging all elements in $E^{(k-1)}$ which contain $x_k$ to form a new element $E_k$;

(2)  deleting those elements in $E^{(k-1)}$ which contain $x_k$;

(3)  adding $E_k$.

As an example, consider Gaussian elimination as applied to our nine-point model problem with the nested dissection ordering on a 3 x 3 grid (see Figure 3; the initial elements corresponding to the nonzero entries of A are not shown). As each variable is eliminated, one new element is created and the elements merged to form this element are deleted. Since A is irreducible, the final element contains all the variables in the grid. Note that some of the elements are equal.

The order in which the variables are eliminated determines the elements which are created during the elimination process. We can describe the element merges that take place by an element merge tree:[2]

(1)  The nodes of the tree are the elements which were created during the elimination process (however, if two elements are equal, then they are identified with

_____

[2] The element merge tree is actually a forest unless the matrix A is irreducible.

the same node).

$$x_1 - x_5 - x_2$$
$$| \qquad | \qquad |$$
$$x_7 - x_8 - x_9$$
$$| \qquad | \qquad |$$
$$x_3 - x_6 - x_4$$

$E^{(0)} = \{ \ldots \}$

$E^{(1)} = \{ E_1=\{x_1,x_5,x_7,x_8\}, \ldots \}$

$E^{(2)} = \{ E_1, E_2=\{x_2,x_5,x_8,x_9\}, \ldots \}$

$E^{(3)} = \{ E_1, E_2, E_3=\{x_3,x_6,x_7,x_8\}, \ldots \}$

$E^{(4)} = \{ E_1, E_2, E_3, E_4=\{x_4,x_6,x_8,x_9\}, \ldots \}$

$E^{(5)} = \{ E_3, E_4, E_5=\{x_1,x_2,x_5,x_7,x_8,x_9\}, \ldots \}$

$E^{(6)} = \{ E_5, E_6=\{x_3,x_4,x_6,x_7,x_8,x_9\}, \ldots \}$

$E^{(7)} = \{ E_7=\{x_1,x_2,x_3,x_4,x_5,x_6,x_7,x_8,x_9\}, \ldots \}$

$E^{(8)} = \{ E_8=E_7, \ldots \}$

$E^{(9)} = \{ E_9=E_7 \}$

Figure 3

(2) A node (i.e., element) $E_i$ is a son of another node $E_j$ if and only if $E_i \neq E_j$ and $E_i$ was merged to form $E_j$ when variable $x_j$ was eliminated.

The merge tree for the previous example appears in Figure 4.

Note the following properties of element merge trees:

(1) Every variable $v_k$ corresponds to some node in the element merge tree, namely the node identified with $E_k$; however, several variables will correspond to the

same node in the tree if the corresponding elements
are equal.

$$x_1 - x_5 - x_2$$
$$| \quad | \quad |$$
$$x_7 - x_8 - x_9$$
$$| \quad | \quad |$$
$$x_3 - x_6 - x_4$$

$$\{x_1, x_2, x_3, x_4, x_5, x_6, x_7, x_8, x_9\} = E_7 = E_8 = E_9$$

$$\{x_1, x_2, x_5, x_7, x_8, x_9\} \ E_5 \qquad \{x_3, x_4, x_6, x_7, x_8, x_9\} = E_6$$

$$\{x_1, x_5, x_7, x_8\} \quad \{x_2, x_5, x_8, x_9\} \quad \{x_3, x_6, x_7, x_8\} \quad \{x_4, x_6, x_8, x_9\}$$
$$= E_1 \qquad\qquad = E_2 \qquad\qquad = E_3 \qquad\qquad = E_4$$

Figure 4

(2)  If values were known for all the variables $x_i$ whose
     corresponding $E_i$ is equal to the root element of the
     tree, then the problem would split into subproblem(s),
     which could be solved in the same fashion.

This allows us to generalize the "Disaster Strikes"
Algorithm to non-model problems.

Minimal Storage Sparse Elimination (MSSE) Algorithm:

(1)  Construct the element merge tree corresponding to the
     given order of elimination.

(2)  Solve for the variables corresponding to the root
     element in the tree.

(3)   The problem splits; solve each of the subproblems in the same fashion.

Moreover, there is no reason why we cannot solve for more than just the variables corresponding to the root node at each stage.   Instead, we could solve for as many variables as we have storage for the necessary nonzero entries in the reduced system.   The result will be less work, although, since the bulk of the work is in the first one or two stages, the savings will be primarily in bookkeeping operations rather than in actual arithmetic.   But how do we implement such a scheme?

4.   An Implementation of MSSE.   In this section, we present an alternate formulation of Gaussian elimination [6], and show how it leads to an implementation of the MSSE algorithm similar to the assembly approach for solving finite-element equations [9].

Gaussian elimination can be expressed as a sequence of operations on the coefficients and right-hand sides of the original system of equations leading to the reduced system:

SET   $a_{ij}^{(0)} = a_{ij}$,   $b_i^{(0)} = b_i$   $1 \le i \le j \le N$

FOR   k = 1, 2, ..., N-1   DO

$$
a_{ij}^{(k)} = \begin{cases} a_{ij}^{(k-1)} - \dfrac{a_{ki}^{(k-1)} \, a_{kj}^{(k-1)}}{a_{kk}^{(k-1)}} & k < i \le j \le N \\[3ex] a_{ij}^{(k-1)} & \text{otherwise} \end{cases}
$$

$$b_i^{(k)} = \begin{cases} b_i^{(k-1)} - \dfrac{a_{ki}^{(k-1)} b_k^{(k-1)}}{a_{kk}^{(k-1)}} & k < i \le N \\[20pt] b_i^{(k-1)} & \text{otherwise} \end{cases}$$

Here $a_{ij}^{(k)}$ and $b_i^{(k)}$ are the entries of the partially reduced system after $x_1, \ldots, x_k$ have been eliminated. The importance of this formulation lies in the following result:

<u>Theorem</u>: If

$$a_{ij}^{(k)} \ne 0, \quad k < i \le j \le N,$$

then there exists an element $E_m$ in $E^{(k)}$ such that

$$x_i, x_j \in E_m, \quad k < i \le j \le N.$$

If we exclude the possibility of exact cancellation, then the converse is also true.

This result suggests how to store the nonzero entries of the partially reduced systems. Associate with each element $E_m \in E^{(k)}$ a matrix $C_m = [c_{x_i, x_j}^{(k)}]$ and a vector $d_m = [d_{x_i}^{(k)}]$ whose rows and columns correspond to those $x_i \in E_m$ with $i > m$; and whose values represent <u>corrections</u> to the entries of A and b resulting from the elimination of those variables $x_i \in E_m$ with $i \le m$. For the nine-point problem on a 3 x 3 grid, the $C_1$, $d_1$ and $C_5$, $d_5$ corresponding to $E_1$ and $E_5$ are shown in Figure 5.

To compute the corrections associated with a new element, we perform an assembly and elimination process similar to that used in finite-element solutions (see

$$x_1 - x_5 - x_2$$
$$| \quad | \quad |$$
$$x_7 - x_8 - x_9$$
$$| \quad | \quad |$$
$$x_3 - x_6 - x_4$$

$$E_1 = \{x_1, x_5, x_7, x_8\}$$

$$c_{x_i, x_j}^{(1)} = - \frac{a_{1i}^{(0)} a_{1j}^{(0)}}{a_{11}^{(0)}} \qquad i \leq j, \quad i,j \in \{5,7,8\}$$

$$d_{x_i}^{(1)} = - \frac{a_{1i}^{(0)} b_1^{(0)}}{a_{11}^{(0)}} \qquad i \in \{5,7,8\}$$

$$E_5 = \{x_1, x_2, x_5, x_7, x_8, x_9\}$$

$$c_{x_i, x_j}^{(5)} = - \frac{a_{1i}^{(0)} a_{1j}^{(0)}}{a_{11}^{(0)}} - \frac{a_{2i}^{(1)} a_{2j}^{(1)}}{a_{22}^{(1)}} - \frac{a_{5i}^{(4)} a_{5j}^{(4)}}{a_{55}^{(4)}}$$

$$i \lessgtr j, \quad i,j \in \{7,8,9\}$$

$$d_{x_i}^{(5)} = - \frac{a_{1i}^{(0)} b_1^{(0)}}{a_{11}^{(0)}} - \frac{a_{2i}^{(1)} b_2^{(1)}}{a_{22}^{(1)}} - \frac{a_{5i}^{(4)} b_5^{(4)}}{a_{55}^{(4)}}$$

$$i \in \{7,8,9\}$$

Figure 5

Figure 6):

(1) Set up a workspace with the rows and columns corresponding to those variables $x_i \in E_k$ with $i \geq k$ and an
additional column corresponding to the right-hand
side; the workspace is initialized to zeroes.

$$\begin{array}{ccccc} x_1 & - & x_5 & - & x_2 \\ | & & | & & | \\ x_7 & - & x_8 & - & x_9 \\ | & & | & & | \\ x_3 & - & x_6 & - & x_4 \end{array} \qquad \begin{array}{l} E_1 = \{x_1,x_5,x_7,x_8\} \\ \\ E_2 = \{x_2,x_5,x_8,x_9\} \\ \\ E_5 = \{x_1,x_2,x_5,x_7,x_8,x_9\} = E_1 \cup E_2 \cup \ldots \end{array}$$

WORKSPACE

| | $x_5$ | $x_7$ | $x_8$ | $x_9$ | b |
|---|---|---|---|---|---|
| $x_5$ | A+1+2 | A+1 | A+1+2 | A+2 | b+1+2 |
| $x_7$ | | 1 | 1 | 0 | 1 |
| $x_8$ | | | 1+2 | 2 | 1+2 |
| $x_9$ | | | | 2 | 2 |
| | | | $C_k$ | | $d_k$ |

A and b denote contributions from an initial element

1 denotes a contribution from $E_1$

2 denotes a contribution from $E_2$

Figure 6

(2) The initial elements merged to form $E_k$ are of the form $\{x_k,x_i\}$ where $a_{ki} \neq 0$; replace the $(x_k,x_i)$ entry in the workspace by $a_{ki}$ and the $(x_k,b)$ entry by $b_k$.

(3) For those elements $E_m$ merged to form $E_k$, add the corrections associated with $E_m$ to the appropriate entries of the workspace.

(4) Eliminate $x_k$; the first row of the workspace is the corresponding row of the reduced system; the remaining rows are the $C_k$ and $d_k$ associated with $E_k$.

If several variables all correspond to the same node in the tree, then we eliminate them simultaneously.

Note that as long as all variables corresponding to descendants of a node are eliminated before variables corresponding to that node, then the "same" operations are done during the elimination process. Thus, instead of eliminating variables in a breadth-first or bottom-up order, we could equally well eliminate them in a depth-first order (see Figure 7). This cuts the amount of

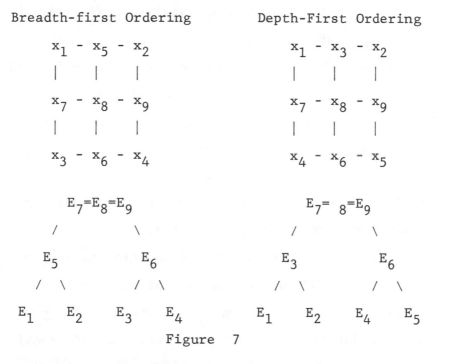

Breadth-first Ordering          Depth-First Ordering

Figure   7

storage required and greatly simplifies storage management [2]. It is straight-forward but tedious to show that

Corollary: For the nine-point model problem with the nested dissection ordering on an $n \times n$ grid, at most $\frac{7}{2}n^2$ nonzero entries have to be stored at any stage of the elimination.

5. Experimental Results. In this section, we present the results of some experiments with a code [2] which implements the Minimal Storage Sparse Elimination algorithm, and discuss the advantages and disadvantages of such a code versus a good sparse elimination code.

Tests were run on the nine-point model problem with the nested dissection ordering on a 31 x 31 grid. As a base for comparison, the same problem was solved using the symmetric codes from the Yale Sparse Matrix Package (YSMP) [5]. The times[3] are presented in Table 1.

In the table, Full Recursion refers to solving for only the variables corresponding to the root element at each stage; No Recursion refers to saving the entire reduced system in core, and solving for all the variables in the first stage; and Partial Recursion refers to saving as much of the reduced system as possible at each stage, and solving for the corresponding variables.

As the times show, the MSSE implementation is nearly competitive with YSMP when the amount of storage is sufficiently large, but less so when the amount of storage is reduced. The differences arise both from the additional bookkeeping required to implement sparse elimination in this manner and the additional work required to recompute discarded values.

---

[3] All experiments were run on a DEC KL-20 computer with the FORTRAN-20 optimizing compiler.

Table 1:  Nine-point operator on a 31 x 31 grid

|  | Time (seconds) | Additional Storage |
|---|---|---|
| YSMP |  |  |
| Preprocessing (symbolic factorization) | 0.309 |  |
| Numerical factorization and solution | 3.291 |  |
| Total | 3.600 | 29198 |
| MSSE (Full Recursion) |  |  |
| Preprocessing (element merge tree) | 0.547 |  |
| Numerical factorization and solution | 10.624 |  |
| Total | 11.171 | 7426 |
| MSSE (No Recursion) |  |  |
| Numerical factorization and solution | 4.178 |  |
| Total | 4.725 | 32169 |
| MSSE (Partial Recursion) |  |  |
| Numerical factorization and solution | 6.814 |  |
| Total | 7.361 | 8124 |

The principal advantages of the YSMP code are speed and
the ability to solve efficiently additional systems with
the same coefficient matrix but different right-hand sides
(since the entire reduced system is retained).  The principal advantages of MSSE are the ability to solve problems
in significantly less core, and to trade off an increase
in execution time for a decrease in core.

6.  Applications to Auxiliary Storage Sparse Elimination.
In this section, we examine how the ideas underlying MSSE

can be applied to auxiliary storage sparse elimination.

Auxiliary storage devices are characterized by long access times and high transfer rates. Therefore it is important to transfer as many words as possible as infrequently as possible, consistent with carrying out the elimination process. Unfortunately, neither of the standard approaches to sparse elimination -- the row-by-row approach [5] and the outer-product approach [1] -- allows auxiliary storage to be integrated in this manner. Both would be characterized by excessive input/output to random-access storage. Thus input/output could easily dominate computation in the total run-time. However, by using the fundamental ideas of MSSE, we can avoid this problem.

Assume that there is enough storage to perform the MSSE algorithm. Recall that MSSE must generate the entire reduced system while solving for the set of variables corresponding to the root element. Clearly it suffices to output these values to auxiliary storage instead of throwing them away, and then read them back in during the backsolution process. Since each value is read and written once, the minimal amount of input/output is performed and, by appropriate buffering, this can be done efficiently.

But what if the amount of storage is insufficient to carry out MSSE? It seems likely that interposing a paging scheme to simulate a larger amount of available core would be a promising approach.

## REFERENCES

[1]  I. S. DUFF, N. MUNKSGAARD, H. B. NIELSON, and
J. K. REID, Direct solution of sets of linear equations whose matrix is sparse, symmetric and indefinite,
Harwell Report CSS-44, January 1977.

[2]  S. C. EISENSTAT, A minimal storage sparse elimination
code.  To appear.

[3]  S. C. EISENSTAT, M. H. SCHULTZ, and A. H. SHERMAN,
Applications of an element model for Gaussian elimination, in J. R. BUNCH and D. J. ROSE, editors,
Proceedings of the Symposium on Sparse Matrix Computations, Argonne National Laboratory (September 1975),
pp. 85-96.

[4]  S. C. EISENSTAT, M. H. SCHULTZ, and A. H. SHERMAN,
Minimal storage band elimination, in A. H. SAMEH and
D. KUCK, editors, Proceedings of the Symposium on
High Speed Computer and Algorithm Organization, University of Illinois at Champagne-Urbana (April 1977),
pp. 273-286.

[5]  S. C. EISENSTAT, M. C. GURSKY, M. H. SCHULTZ, and
A. H. SHERMAN, Yale sparse matrix package:  I.  The
symmetric codes, Yale Computer Science Research Report #112, July 1977.

[6]  G. E. FORSYTHE and C. B. MOLER, Computer Solution of
Linear Algebraic Systems, Prentice-Hall, Englewood
Cliffs, N.J., 1967.

[7]  J. A. GEORGE, Nested dissection of a regular finite
element mesh, SIAM Journal on Numerical Analysis 10
(1973), pp. 345-363.

[8]  A. J. HOFFMAN, M. S. MARTIN, and D. J. ROSE, Complexity bounds for regular finite difference and
finite element grids, SIAM Journal on Numerical
Analysis 10 (1973), pp. 364-369.

[9]  B. M. IRONS, A frontal solution program for finite
elements, International Journal for Numerical Methods
in Engineering 2 (1970) pp. 5-32.

[10]  D. J. ROSE and G. F. WHITTEN, Automatic nested
dissection, Proceedings of the 1974 ACM National
Conference, pp. 82-88.

[11]  J. H. WILKINSON, The Algebraic Eigenvalue Problem,
Clarendon Press, Oxford, 1965.

# A Quotient Graph Model
# for Symmetric Factorization†

## Alan George‡ and Joseph W. H. Liu*

Abstract. Consider the computation of the Cholesky factorization $LL^T$ of the symmetric positive definite matrix A. In this paper we present a model of this factorization algorithm based on quotient graphs, and discuss its relationship to some existing models. The primary advantage of our model is that its computer implementation is very efficient. In particular, we show how it can be implemented in space proportional to the number of nonzeros in A. Some numerical experiments showing the performance of an implementation of the minimum degree algorithm using the model are provided, along with some comparisons with implementations based on other models.

1. Introduction. Consider the symmetric positive definite system of linear equations

(1.1)        Ax = b ,

where the N by N coefficient matrix A is sparse. If Cholesky's method is to be used to solve (1.1), there are compelling reasons to proceed in four distinct steps as follows:

Step 1 (Ordering Step)  Find a "good" ordering for the system, and transform it, perhaps only implicitly, to

(1.2)        $(PAP^T)(Px) = Pb.$

Step 2 (Symbolic Factorization)  Determine the location of the nonzeros in L, where $PAP^T = LL^T$, and set up the appropriate data structures for L.

† Research supported in part by Canadian National Research Council grant A8111.

†† Department of Computer Science, University of Waterloo, Waterloo, Ontario, Canada, N2L 3G1.

* Systems Dimensions Limited, 111 Avenue Road, Toronto, Ontario, Canada.

Step 3   (Numerical Factorization)  Decompose  $PAP^T = LL^T$,
          using Cholesky's method.

Step 4   (Triangular Solution)  Solve  $Ly = Pb$, $L^Tz = y$, and
          then set  $x = P^Tz$.

Our objective in this paper is to present a model of
symmetric factorization which is useful in the analysis and
implementation of Steps 1 and 2.  Sparse matrices normally
suffer some fill-in when they are factored, so that  $L + L^T$
has nonzeros in positions which are zero in  A.  The objective of Step 1 is usually to find a permutation  P  so
that this fill-in is acceptably low.  Several ordering
algorithms, such as the minimum degree algorithm [4,8],
essentially require that the factorization be simulated as
the ordering proceeds, because ordering decisions depend
upon the structure of the partially factored matrix.  Thus,
it is crucial to be able to simulate the factorization
efficiently.  Similarly, the implementation of Step 2 above
also depends upon being able to simulate the factorization
efficiently, in terms of both space and time.

An outline of the paper is as follows.  In Section 2 we
review some existing models of symmetric elimination (factorization), and discuss some of their important features.
In Section 3 we present our quotient graph model, and compare it to those of Section 2.  In Section 4 we show that
the space required for a computer implementation of the
model is proportional to the number of nonzeros in  A
(that is, independent of the fill-in).  Section 5 contains
some numerical experiments comparing the performance of
three implementations of the minimum degree algorithm, which
use data structures based on different models of the elimination process.  The experiments support our contention
that our quotient graph model has important practical
applications.  Section 6 contains our concluding remarks.

## 2.  Some Existing Models  or Characterizations of Symmetric Factorization

### 2.1  Matrix Formulation

As a point of departure, we describe the factorization in terms of the actual numerical computation that is performed.  The models we subsequently consider are used to simulate what happens in terms of structural changes to the matrices, without actually involving any numerical computation.  Setting $A = A_0 = H_0$, the factorization of  A  can be described by the following equations.

$$A_0 = \begin{pmatrix} d_1 & v_1^T \\ v_1 & \overline{H}_1 \end{pmatrix}$$

$$= \begin{pmatrix} \sqrt{d_1} & 0 \\ v_1/\sqrt{d_1} & I_{N-1} \end{pmatrix} \begin{pmatrix} 1 & 0 \\ 0 & \overline{H}_1 - v_1 v_1^T/d_1 \end{pmatrix} \begin{pmatrix} \sqrt{d_1} & v_1^T/\sqrt{d_1} \\ 0 & I_{N-1} \end{pmatrix} = L_1 A_1 L_1^T$$

$$A_1 = \begin{pmatrix} 1 & & 0 \\ & d_2 & & v_2^T \\ 0 & & \\ & v_2 & & \overline{H}_2 \end{pmatrix}$$

$$\begin{pmatrix} 1 & & 0 \\ & \sqrt{d_2} & & 0 \\ 0 & & \\ & v_2/\sqrt{d_2} & I_{N-2} \end{pmatrix} \begin{pmatrix} 1 & & 0 \\ & 1 & & 0 \\ 0 & & \\ & 0 & H_2 \end{pmatrix} \begin{pmatrix} 1 & & 0 \\ & \sqrt{d_2} & & v_2^T/\sqrt{d_2} \\ 0 & & \\ & 0 & I_{N-2} \end{pmatrix}$$

$$\vdots$$

$$A_n = L_N I_N L_N^T.$$

It is straightforward to show that  $A = LL^T$, where

$$L = \sum_{k=1}^{N} L_k - (N-1) I_N .$$

Here $I_k$ denotes a $k$ by $k$ identity matrix, $d_k$ is a positive scalar, and $v_k$ is a vector of length N-k. The matrix $H_k$ is an N-k by N-k symmetric positive definite matrix, which we refer to as "the part of A remaining to be factored after the first k steps of the factorization have been performed".

## 2.2 Some Elementary Graph Theory Terminology

The models we discuss in subsequent parts of this paper rely heavily on graph theory terminology, so in this section we introduce some of the essential definitions and notions. For our purposes, a <u>graph</u> G = (X,E) consists of a finite nonempty set X of <u>nodes</u> together with a prescribed <u>edge set</u> E of unordered pairs of distinct nodes. A graph G' = (X',E') is a <u>subgraph</u> of G if X' ⊂ X and E' ⊂ E. For Y ⊂ X, G(Y) refers to the subgraph (Y,E(Y)) of G, where E(Y) = {{ u,v} ∈ E|u,v ∈ Y}.

Nodes x and y of X are <u>adjacent</u> if { x,y} ∈ E. For Y ⊂ X, the set of nodes adjacent to Y is defined and denoted by

$$\text{Adj}(Y) = \{ x \in X-Y | \{ x,y\} \in E \text{ for some } y \in Y\} .$$

If Y = { y}, we write Adj(y) rather than Adj({ y}). The <u>degree</u> of a node x is simply the number of nodes adjacent to it, denoted by deg(x). We refer to y ∈ Adj(x) as a <u>neighbor</u> of x. A set of nodes Y ⊂ X which are pairwise adjacent is a <u>clique</u>.

A <u>path</u> of <u>length</u> ℓ is an ordered set of distinct vertices $(v_0, v_1, \ldots, v_\ell)$ where $v_i \in \text{Adj}(v_{i-1})$ for $1 \le i \le \ell$. A graph G is <u>connected</u> if there is a path connecting each pair of distinct nodes. If G is disconnected, it consists of two or more maximal connected <u>components</u>.

Let S ⊂ X, and w ∈ X-S. The node w is said to be <u>reachable</u> from y <u>through</u> S if there exists a path $(y, y_1, y_2, \ldots, y_k, w)$ such that $y_i \in S$, for $1 \le i \le k$. We allow k to be zero, so any node w ∈ X-S and adjacent to y is reachable from y. The <u>reach set of y through S</u> is then denoted and defined by

Reach$(y,S)$ = { w $\in$ X-S|w is reachable from y through S}.

We extend this definition to subsets  Y  as follows.  Let
Y $\subset$ X  and  Y $\cap$ S = $\phi$.  The reach set of Y through S is then
Reach$(Y,S)$ = { w $\in$ X-(S$\cup$ Y)|w is reachable from some node
y $\in$ Y through S} .

Note that when  S = $\phi$, Reach$(Y,S)$ = Adj$(Y)$.

In subsequent sections of this paper, we will be applying
these definitions to various graphs.  When the graph being
referred to is not absolutely clear from context, we will
put the appropriate subscript on the definition.  Thus,
notations of the following type will be used:  Adj$_G(Y)$,
deg$_G(Y)$, Reach$_G(Y,S)$, etc.

### 2.3  Elimination Graph Model

In this section we describe the graph theory approach to
symmetric elimination which was introduced by Parter [7],
and popularized and extended by Rose [8].  Let  A  be an
N  by  N  symmetric matrix.  The labelled undirected graph
of  A, denoted by  $G^A = (X^A, E^A)$, is one for which  $X^A$  is
labelled from 1 to  N:

$$X^A = \{ x_1, x_2, \ldots, x_N \} ,$$

and { $x_i, x_j$ } $\in$ $E^A$ if and only if  $A_{ij} \neq 0$.  For any  N  by
N  permutation matrix  P, the unlabelled graphs of  A  and
PAP$^T$ are the same, but the associated labellings differ.
Thus, the graph of  A  is a convenient vehicle for studying
the structure of  A, since no particular ordering is im-
plied by the graph.

Now consider the symmetric factorization of  A  into LL$^T$
using the algorithm described in Section 2.1.  The sparsity
changes (fill-in) can be modelled by a sequence of graph
transformations on  $G^A$.  Let  G = (X,E)  be a graph and
y $\in$ X.  The elimination graph of  G  by  y, denoted by  $G_y$,
is the graph  (X - { y }, E(X - { y })$\cup${{ u,v}|u,v $\in$ Adj$(y)$}).
In words,  $G_y$  is obtained from  G  by deleting  y  and its
incident edges, and then adding edges to the remaining

graph so that the set $\text{Adj}(y)$ is a clique. This recipe
is due to Parter [7].

With this definition, the process of symmetric elimination on $A$ can be modelled as a sequence of elimination
graphs $G_0, G_1, G_2, \ldots, G_{N-1}$, where $G_0 = G^{\bar{A}}$ and $G_i = (G_{i-1})_{x_i}$
$= (X_i, E_i)$, $i = 1, 2, \ldots, N-1$. Here $X_i = \{ x_{i+1}, x_{i+2}, \ldots, x_N \}$,
and it is straightforward to verify that $G_i \equiv G^{H_i}$, where
$H_i$ (defined in Section 2.1) is the part of the matrix
remaining to be factored after step $i$ of the factorization has been completed. Thus, this model is quite explicit; the structure of $G_i$ corresponds directly to the
matrix $H_i$.

### 2.4 Characterization of Symmetric Elimination in Terms of Reachable Sets

Again let $G^A = (X^A, E^A)$ be the graph of $A$, and let
$G^F = (X^F, E^F)$ be the graph of $F = L+L^T$. We call $G^F$ the
filled graph of $G^A$, where $X^F = X^A$ and $E^F$ consists of
$E^A$ together with those edges added during the factorization.
The relationship between $E^A$ and $E^F$ is contained in the
following lemma due to Parter [7].

**Lemma 2.2** The unordered pair $\{ x_i, x_j \} \in E^F$ if and only
if $\{ x_i, x_j \} \in E^A$, or there exists some $k < \min\{ i, j \}$ for
which $\{ x_i, x_k \} \in E^F$ and $\{ x_j, x_k \} \in E^F$. $\quad\square$

The above lemma is somewhat unsatisfactory because it is
recursive in $E^F$. The following result, which is essentially a restatement in our notation of a lemma due to
Rose et. al. [9], relates $E^F$ directly to $E^A$. Let
$S_i = \{ x_1, x_2, \ldots, x_i \}$, $i = 1, \ldots, N$, with $S_0 = \phi$.

**Lemma 2.3 [4]** Let $j > i$. Then the unordered pair
$\{ x_i, x_j \} \in E^F$ if and only if $x_j \in \text{Reach}_G(x_i, S_{i-1})$. $\quad\square$

Thus, the sets $\{ S_i \}$ together with the Reach operator
precisely characterize the adjacency structure of the
elimination graphs $\{ G_i \}$. In particular, we have

**Lemma 2.4** [4]    Let   y   be a node in the elimination graph
$G_i = (X_i, E_i)$.    The set of nodes adjacent to   y   in   $G_i$   is
given by   $Reach_G(y, S_i)$.                                                    □

Note that in Lemma 2.3, and Lemma 2.4, the Reach operator is applied to the original graph   G.

### 3.    The Quotient Graph Model

### 3.1    Notation and Definitions

Let   G = (X,E)   be a graph with   X   the set of nodes and
E   the set of edges.    For a subset   S ⊂ X, recall that G(S) is
used to refer to the <u>subgraph</u>   (S,E(S))   of   G, where
E(S) = {{ u,v} ∈ E|u,v ∈ S}.    The central notion in the new
model is that of a quotient graph [3], which we now review.
Let   $P$   be a partitioning of the node set   X:

$$P = \{ Y_1, Y_2, \ldots, Y_p \} .$$

That is,   $\bigcup_{k=1}^{p} Y_k = X$   and   $Y_i \cap Y_j = \phi$   for   i ≠ j.    The
<u>quotient graph</u> of   G   with respect to   $P$   is defined to be
the graph   $(P, \mathcal{E})$, where   $\{ Y_i, Y_j \} \in \mathcal{E}$   if and only if
$Y_i \cap Adj(Y_j) \neq \phi$.    This graph will be denoted by   G/$P$.
Consider the example in Figure 3.1.    If
$P = \{\{ a,b,c\}, \{ d,e\}, \{ g\}, \{ f,h\}\}$   is the partitioning, the
quotient graph   G/$P$   is given as shown.

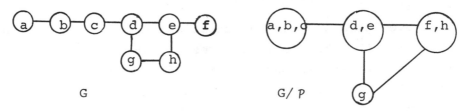

G                                    G/ $P$

Figure 3.1   A quotient graph

An important type of partitioning is that defined by
connected components.    Let   S   be a subset of the node set
X.    The <u>component partitioning</u>   $C(S)$   of   S   is defined as

$C(S) = \{ Y ⊂ S|G(Y)$   is a connected component in the sub-
graph   G(S)}.

When $S = X$, $C(X)$ simply contains the component sets in the graph $G$. Therefore, the corresponding quotient graph $G/C(X)$ consists of $|C(X)|$ isolated nodes.

The structure of the quotient graph $G/C(X)$ is not particularly interesting. However, we now study a closely related type of partitioning, which turns out to be quite relevant in the modelling of Gaussian elimination. Again let $S$ be a subset of $X$. The <u>partitioning</u> on $X$ <u>induced</u> <u>by</u> the subset $S$ is defined to be $\overline{C}(S) = C(S) \cup \{\{x\} | x \notin S\}$. That is, the partitioning $\overline{C}(S)$ consists of the component partitioning of $S$ and the remaining nodes of the graph $G$.

Consider the graph in Figure 3.1. Let $S$ be the subset $\{a,b,d,f,g\}$. It can be seen that $C(S) = \{\{a,b\}, \{d,g\},\{f\}\}$, so that $\overline{C}(S)$ has six members and is given by $\overline{C}(S) = \{\{a,b\}, \{d,g\}, \{f\}, \{c\}, \{e\}, \{h\}\}$. In this case, the quotient graph $G/\overline{C}(S)$ is given in Figure 3.2. Here, we use double circles to indicate those partition members in $C(S)$.

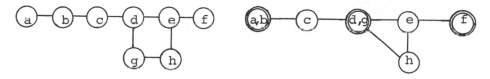

Figure 3.2  The quotient graph  $G/\overline{C}(S)$

## 3.2  The Model

In Section 2, we have reviewed some existing models of symmetric elimination. These models are used to study the Gaussian elimination process. In this section, we intro-duce a new model using the notion of quotient graphs. Its relationship with those in Section 2 will also be discussed.

Consider the symmetric factorization of a matrix $A$ into $LL^T$. Recall from Section 2 that the elimination pro-cess applied to $A$ can be interpreted as a sequence of elimination graphs $G_0, G_1, \ldots, G_{N-1}$. The graph $G_i$ pre-cisely reflects the structure of the matrix remaining to be factored after the i-th step of the Gaussian elimination.

The new model represents the process as a sequence of quotient graphs, which may be regarded as <u>implicit repre-sentations</u> of the elimination graphs $\{ G_i \}$. Let $G = (X,E)$ be the graph and $x_1, x_2, \ldots, x_N$ be the sequence of node elimination.

As in Section 2.4, let $S_i = \{ x_1, \ldots, x_i \}$ for $1 \le i \le N$ and $S_0 = \phi$. Consider the partitioning $\bar{C}(S_i)$ induced by $S_i$, and the corresponding quotient graph $G/\bar{C}(S_i)$. We shall denote this quotient graph by $G_i$. In this way, we obtain a sequence of quotient graphs $G_1, G_2, \ldots, G_N$. Figure 3.3 contains an example. Partition members in $C(S_i)$ are marked in double circles.

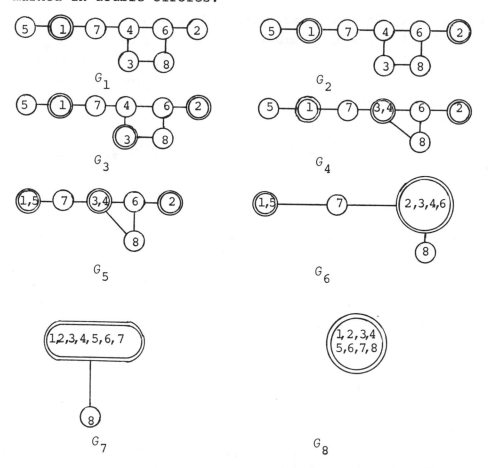

Figure 3.3  A sequence of quotient graphs

As we stated earlier, the quotient graph $G_i = G/\overline{C}(S_i)$ can be regarded as an implicit representation of the elimination graph $G_i$; they are related by the following result.

Lemma 3.1 For $y \notin S_i$,

$$\text{Reach}_{G_i}(\{y\}, C(S_i)) = \{\{x\} \mid x \in \text{Reach}_G(y, S_i)\}.$$

Proof Consider $x \in \text{Reach}_G(y, S_i)$. If $x$ and $y$ are adjacent in $G$, so are $\{x\}$ and $\{y\}$ in $G_i$. Otherwise, there exists a path $y, s_1, \ldots, s_t, x$ in $G$ where $\{s_1, \ldots, s_t\} \subset S_i$. Let $C$ be the component in $G(S_i)$ containing $\{s_1, \ldots, s_t\}$. Then we have a path $\{y\}, C, \{x\}$ in $G_i$ so that $\{x\} \in \text{Reach}_{G_i}(\{y\}, C(S_i))$.

Conversely, consider any $\{x\} \in \text{Reach}_{G_i}(\{y\}, C(S_i))$. There exists a path $\{y\}, C_1, \ldots, C_t, \{x\}$ in $G_i$ where each $C_j \in C(S_i)$. If $t = 0$, then $x$ and $y$ are adjacent in the original graph $G$. If $t > 0$, by the definition of $C(S_i)$, $t$ cannot be greater than one; that is, the path must be $\{y\}, C, \{x\}$. Since $G(C)$ is a connected subgraph, we can obtain a path from $y$ to $x$ through $C \subset S_i$ in the graph $G$. Hence $x \in \text{Reach}_G(y, S_i)$. □

We can obtain the __structure__ of the elimination graph $G_i$ from $G_i$ as follows.

Step 1 Remove nodes in $C(S_i)$ and their incident edges from the quotient graph $G_i$.

Step 2 For each $C \in C(S_i)$, add edges to the quotient graph so that all adjacent nodes of $C$ form a clique in the elimination graph.

To illustrate the idea, consider the transformation of $G_4$ to $G_4$ for the example in Figure 3.3. The elimination graph $G_4$ is given in Figure 3.4.

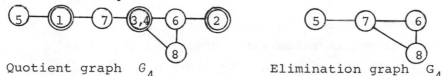

Quotient graph $G_4$        Elimination graph $G_4$

Figure 3.4 From quotient graph to elimination graph.

In terms of implicitness, the quotient graph model lies in between the reachable set model and the elimination graph model, as a vehicle for simulating the elimination process.

Reachable set
on original $\longrightarrow$ Quotient $\longrightarrow$ Elimination
graph                graph               graph

Since it is more explicit than the reachable set model in the representation, less effort is usually required to produce the adjacency sets for the elimination graph. On the other hand, the new model has the advantage over the explicit elimination graph model in that it requires a fixed amount of storage in its computer implementation. This point will be elaborated upon in the next section.

## 4. Computer Implementation of the Quotient Graph Model

## 4.1 Preliminary Results

In this subsection, some simple but important properties of the quotient graph model will be established. These will be used to show that the model can be implemented in-place. We discuss the implementation in Section 4.2. Let $G = (X,E)$ be a given graph.

**Lemma 4.1** Let $S \subset X$ where $G(S)$ is a connected sub-graph. Then

$$\sum_{x \in S} |Adj(x)| \geq |Adj(S)| + 2(|S| - 1) .$$

**Proof** Since $G(S)$ is connected, there are at least $|S| - 1$ edges in the subgraph. These edges are counted twice in $\sum_{x \in S} |Adj(x)|$ and hence the result. $\square$

We now show that the edge set sizes of the quotient graphs $G_i$ cannot increase with increasing $i$. Let $x_1, x_2, \ldots, x_N$ be the node sequence, let $G_i = (X_i, E_i)$, $0 \leq i < N$, be the corresponding elimination graph sequence, and let $G_i = (\overline{C}(S_i), \mathcal{E}_i)$, $1 \le i \le N$, be the corresponding quotient graph sequence.

**Theorem 4.2** For $1 \le i \le N$, $|\overline{C}(S_{i+1})| \le |\overline{C}(S_i)|$ and
$$|\mathcal{E}_{i+1}| \le |\mathcal{E}_i|.$$

**Proof** Since $C(S_i)$ is the set of components in the subgraph $G(S_i)$, we have $|C(S_{i+1})| \le |C(S_i)| + 1$. However, $|\overline{C}(S_i)| = |C(S_i)| + N - i$, so that

$$\begin{aligned}
|\overline{C}(S_{i+1})| &= |C(S_{i+1})| + N - i - 1 \\
&\le |C(S_i)| + N - i \\
&= |\overline{C}(S_i)|.
\end{aligned}$$

For the inequality on the edge size, consider $C(S_{i+1})$. If $\{x_{i+1}\} \in C(S_{i+1})$, clearly $|\mathcal{E}_{i+1}| = |\mathcal{E}_i|$. Otherwise, the node $x_{i+1}$ is merged with some components in $C(S_i)$ to form a new component in $C(S_{i+1})$. But then Lemma 4.1 applies, so that $|\mathcal{E}_{i+1}| < |\mathcal{E}_i|$. Hence in all cases, $|\mathcal{E}_{i+1}| \le |\mathcal{E}_i|$. □

The next theorem shows that the degrees of the nodes in the quotient graphs also decrease monotonically with $i$. The proof follows easily from the definition of the quotient graphs $\{G_i\}$ and is omitted.

**Theorem 4.3** For $x \notin S_{i+1}$,
$$|\text{Adj}_{G_{i+1}}(\{x\})| \le |\text{Adj}_{G_i}(\{x\})|. \qquad □$$

The next theorem illustrates the advantage of the quotient graph model over the elimination graph model, and is one of the primary motivations for our introduction of the model.

**Theorem 4.4** $\displaystyle \max_{1 \le i \le N} |\mathcal{E}_i| \le |E| \le \max_{0 \le i < N} |E_i|.$

**Proof** The first inequality follows from Theorem 4.2, and the fact that $|E_0| = |E|$ implies the second one.

□

To illustrate the possible difference between the quantities $\max|\mathcal{E}_i|$ and $\max|E_i|$, we consider the example in Figure 4.1. The corresponding elimination graph and quotient graph sequences are given in Figure 4.2. If we generalize the example in an N-node graph, we have $\max|\mathcal{E}_i| = N - 1$ and $\max|E_i| = (N - 1)(N - 2)/2$.

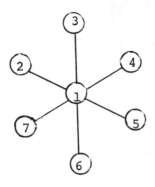

Figure 4.1   A seven-node star graph.

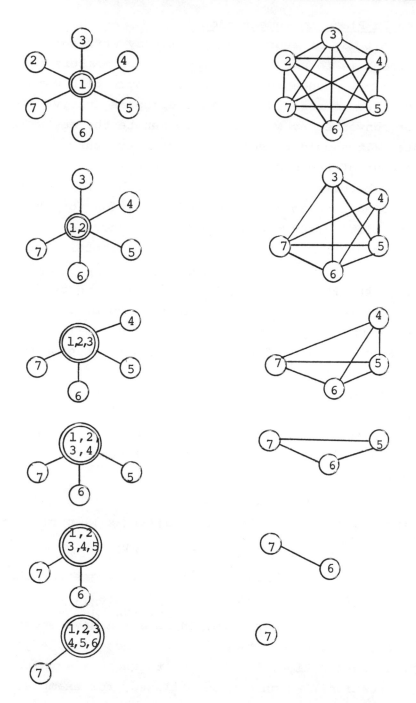

Figure 4.2    The quotient graph and elimination graph
sequences for the example in Figure 4.1.

## 4.2  An In-Place Implementation

A practical way of storing the adjacency structure of a quotient graph is to use <u>representatives</u> for members of the partitioning $P$. Conceptually, for each partition member $C \in P$, we select a node $x \in C$ as the representative for $C$, and for convenience, we use $x$ to denote the member it represents. It should be noted that the structure of $G/P$ is independent of the choice of representatives.

In this subsection, we will demonstrate how we can use Lemma 4.1 to achieve an in-place implementation for the sequence of elimination graphs. We first describe the quotient graph transformation in a more general setting.

Let $G = (X,E)$. Consider a subset $Y \subset X$ and a node $x \notin Y$ such that the subgraph $G(Y \cup \{x\})$ is connected. The problem is to generate the quotient graph by collapsing the subset $Y \cup \{x\}$ to form a new "super-node". The following is a description of an in-place implementation of the transformation.

Step 1  (Form new adjacent set) Determine the set
$\quad\quad$ Adj$(Y \cup \{x\})$.

Step 2  (In-place implementation) Use the node $x$ as the representative of the new quotient member $Y \cup \{x\}$. Reset

$$\text{Adj}(x) \leftarrow \text{Adj}(Y \cup \{x\}).$$

Step 3  (Neighbor update) For $z \in \text{Adj}(Y \cup \{x\})$, put

$$\text{Adj}(z) \leftarrow (\text{Adj}(z) - Y) \cup \{x\}.$$

In Step 2, although $|\text{Adj}(Y \cup \{x\})|$ may be greater than $|\text{Adj}(x)|$, we have by Lemma 4.1 that $\sum_{y \in Y \cup \{x\}} |\text{Adj}(y)| \geq |\text{Adj}(Y \cup \{x\})| + 2|Y|$. Therefore, there are always enough storage locations for $\text{Adj}(Y \cup \{x\})$ from those for $\text{Adj}(y)$, $y \in Y \cup \{x\}$. In addition, for $Y \neq \phi$, there is a surplus of $2|Y|$ locations, which can be utilized, for example, to store links or pointers.

It should also be noted that in Step 3, by Theorem 4.3, the new neighbor set $(\text{Adj}(z) - Y) \cup \{x\}$ in the neighbor update step can also be done in place.

In modelling elimination by the sequence of quotient graphs $\{G_i\}$, the graph $G_{i+1}$ can be obtained from $G_i$ by the execution of the above transformation. More specifically, we use the transformation to collapse the node $x_{i+1}$ with those $C \in C(S_i)$ for which $x_{i+1} \in \text{Adj}(C)$.

To provide a concrete example to demonstrate an in-place implementation, we consider the graph of Figure 3.3 and we assume the adjacency structure is represented as shown in Figure 4.3.

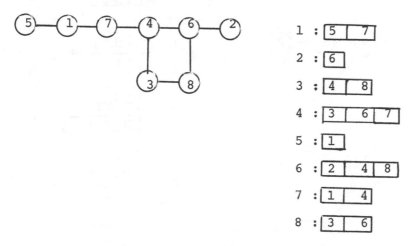

Figure 4.3  A graph and its representation.

Figure 4.4 shows some important steps in producing quotient graphs for this example.  The adjacency structure remains unchanged when the quotient graphs $G_1$, $G_2$ and $G_3$ are formed.  To transform $G_3$ to $G_4$, the nodes 3 and 4 are to be collapsed, so that in $G_4$, the new adjacent set of node 4 contains that of the subset $\{3,4\}$ in the original graph, namely $\{6,7,8\}$.  Here, the last location for $\text{Adj}_G(4)$ is used as a link.  Note also that in the neighbor list of node 8, 3 has been changed to 4 in $G_4$ since node 4 becomes the representative of the component subset $\{3,4\}$.

The representations for $G_5$ and $G_6$ in this storage mode are also included in Figure 4.4.

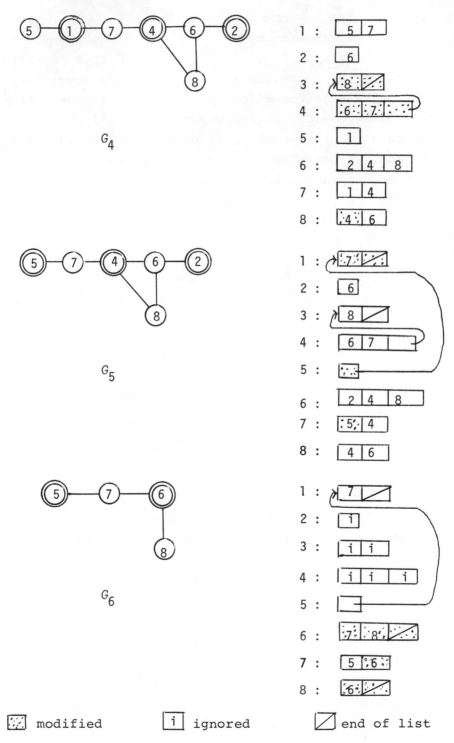

$G_4$

$G_5$

$G_6$

▨ modified      ☐ i ☐ ignored      ▨ end of list

Figure 4.4   An in-place quotient graph transformation.

## 5. An Application of the Model: Implementation of the Minimum Degree Algorithm.

The minimum degree algorithm [4] is a heuristic algorithm which experience has shown to be very effective in finding low-fill orderings for sparse symmetric matrices. The basic idea of the algorithm is, at each elimination step, to permute the part of the matrix remaining to be factored so that a column with the fewest nonzeros is in the pivot position. This implies that at each step of the algorithm we need to know the structure of the partially factored matrix. That is, we need to implement some model of the factorization process.

In this section we provide some experimental results comparing the performance of three implementations of the minimum degree algorithm. The only essential difference among these programs is the way they represent the structure of the partially factored matrix; their data structures are basically efficient implementations of  a) the elimination graph model (MDE) of section 2.3,  b)  the reachable set model discussed in section 2.4 (MDI), and c)  our new quotient graph model (MDQ). The letters in parentheses are the names by which we will refer to the programs, with E, I, and Q intended to imply elimination graph model (explicit), reachable set model (implicit), and quotient graph model.

The implementations and ideas used in these subroutines have been described in detail elsewhere and will not be repeated here. The MDE subroutine is available as part of the Yale Sparse Matrix Package, and is described in [1]. The theory and implementation of the MDI subroutine is documented in [4], and a detailed description of the MDQ code can be found in [6]. However, we do emphasize that these subroutines are basically the same, apart from the way they represent the structure of the matrix. There are some fairly well-known devices for speeding up the algorithm, such as the so-called "mass-elimination" technique [4].

All three implementations employ these devices.

The test problems we chose were the "graded-L" meshes from [3, page 318], which consist of a sequence of similar problems typical of those arising in finite element applications. The results of these runs are summarized in Table 5.1. Execution times are in seconds on an IBM 360/75 computer. The programs are written in Fortran, compiled using the optimizing version of the IBM H-level compiler.

There are numerous noteworthy aspects of the results in Table 5.1. In terms of execution times, the implementation of the minimum degree algorithm based on quotient graphs (MDQ) appears to be substantially faster than either of the other two implementations. Both the MDI and MDQ subroutines appear to execute in time which is very close to proportional to $|E|$, for these problems. The MDE execution time appears to be superlinear in $|E|$, but because of differences in constants of proportionality, it is quite competitive with the MDI code until the size of the problem is several thousand.

Turning now to storage requirements, it is evident that the MDI and MDQ subroutines require storage proportional to $|E|$. It is interesting to note that the actual storage required by the subroutine MDQ is _less_ than that required by MDI. However, MDI does not change the input graph structure, since it generates reachable sets via the original graph. On the other hand, MDQ generates a sequence of graphs from the original, and does not preserve its input. Thus, to be fair in the comparison, we have added space required for the input graph to the MDQ storage requirements. Of course one could preserve the input by writing it on auxiliary storage, and then reading it after the ordering had been found; the apparent storage advantage of MDI over MDQ would then be removed.

The entries in the storage column for MDE were obtained by monitoring the maximum storage used by the working storage arrays. As Figure 4.2 demonstrates, when the eli-

| N | EXECUTION TIME | | | | | | STORAGE REQUIREMENTS | | | | | |
|---|---|---|---|---|---|---|---|---|---|---|---|---|
| | MDE | $\div \lvert E \rvert$ | MDI | $\div \lvert E \rvert$ | MDQ | $\div \lvert E \rvert$ | MDE | $\div \lvert E \rvert$ | MDI | $\div \lvert E \rvert$ | MDQ | $\div \lvert E \rvert$ |
| 265 | .89 | 1.19 | 1.20 | 1.61 | .66 | .887 | 7147 | 9.61 | 4139 | 5.56 | 4832 | 6.49 |
| 406 | 1.44 | 1.25 | 1.97 | 1.71 | 1.14 | .987 | 11151 | 9.65 | 6371 | 5.52 | 7463 | 6.46 |
| 577 | 2.18 | 1.32 | 2.95 | 1.77 | 1.67 | 1.01 | 16735 | 10.11 | 9083 | 5.48 | 10664 | 6.44 |
| 778 | 3.30 | 1.47 | 3.89 | 1.73 | 2.37 | 1.05 | 24697 | 10.99 | 12275 | 5.46 | 14435 | 6.42 |
| 1009 | 4.56 | 1.56 | 5.14 | 1.75 | 2.92 | 1.00 | 32513 | 11.10 | 15947 | 5.45 | 18776 | 6.41 |
| 1270 | 5.76 | 1.56 | 6.30 | 1.72 | 4.16 | 1.12 | 42151 | 11.40 | 20099 | 5.43 | 23687 | 6.40 |
| 1561 | 7.40 | 1.62 | 7.98 | 1.75 | 4.88 | 1.07 | 53215 | 11.67 | 24731 | 5.42 | 29168 | 6.40 |
| 1882 | 9.67 | 1.75 | 9.87 | 1.79 | 6.26 | 1.14 | 65723 | 11.93 | 29843 | 5.41 | 35219 | 6.39 |
| 2233 | 11.55 | 1.76 | 11.98 | 1.83 | 6.88 | 1.05 | 82585 | 12.60 | 35435 | 5.41 | 41840 | 6.39 |
| 2614 | 14.63 | 1.90 | 13.63 | 1.77 | 8.45 | 1.10 | 97121 | 12.64 | 41507 | 5.40 | 49031 | 6.38 |
| | $(\times 10^{-3})$ | $(\times 10^{-3})$ | $(\times 10^{-3})$ | $(\times 10^{-3})$ | $(\times 10^{-3})$ | | | | | | | |

Table 5.1 Performance Statistics on the Graded-L Mesh Problems

mination graphs are stored explicitly as is done by MDE, storage requirements may grow dramatically and then shrink. The disagreeable aspect of this is that the user must estimate the maximum requirement, and provide at least that much to allow the program to execute. The fact that the MDQ and MDI implementations use a fixed predictable and modest amount of storage is a major advantage. As noted before, MDQ implementation enjoys a substantial execution time advantage over the other two.

## 6. Concluding Remarks

As a theoretical tool for analyzing the symmetric factorization process, our model does not appear to be either better than, or inferior to, other models. Its primary advantage, in our opinion, is that it lends itself to very efficient computer implementation. We have given an implementation of the model which requires space only for the original graph of A, and is therefore independent of the fill suffered by A during its factorization. The experiments provided in section 5 demonstrate the practical significance of implementations based on the model.

## 7. References

[1]  S.C. Eisenstat, M.C. Gursky, M.H. Schultz, and A.H. Sherman, The Yale Sparse Matrix Package: I. The Symmetric Codes. Report 112, Dept. of Computer Science, Yale University, 1977.

[2]  Alan George, "Nested dissection of a regular finite element mesh", SIAM J. Numer. Anal. 10 (1973), pp. 345-363.

[3]  Alan George and Joseph W.H. Liu, "Algorithms for matrix partitioning and the numerical solution of finite element systems", SIAM J. Numer. Anal., 15 (1978), pp. 297-327.

[4]  Alan George and Joseph W.H. Liu, "A minimal storage implementation of the minimum degree algorithm", SIAM J. Numer. Anal., to appear.

[5]  Alan George and Joseph W.H. Liu, "An optimal algorithm for symbolic factorization of symmetric matrices", Research Report CS-78-11, Dept. of Computer Science, University of Waterloo, Waterloo, Ont., March 1978.

[6]  Alan George and Joseph W.H. Liu, "A fast implementation of the minimum degree algorithm using quotient graphs", Research Report CS-78-12, Dept. of Computer Science, University of Waterloo, Waterloo, Ontario, May 1978.

[7]  S.V. Parter, "The use of linear graphs in Gauss elimination", SIAM Rev. 3 (1961), pp. 364-369.

[8]  D.J. Rose, "A graph theoretic study of the numerical solution of sparse positive definite systems", in Graph Theory and Computing, R.C. Read, Editor, Academic Press, 1972.

[9]  D.J. Rose, R.E. Tarjan, and G. Lueker, "Algorithmic aspects of vertex elimination on graphs", SIAM J. Comput. 5 (1975), pp. 266-283.

# The Use of Sparse Matrices
# for Image Reconstruction
# from Projections

## Gabor T. Herman*

Abstract. The problem of recovering an image (a function
of two variables) from experimentally available integrals
of its grayness over thin strips is of great importance in
a large number of scientific areas. An important version
of the problem in medicine is that of obtaining the exact
density distribution within the human body from X-ray projections. One approach that has been taken to solve this
problem consists of translating the available information
into a system of linear equalities or inequalities. These
systems are large and sparse, not untypically $10^5$ equations
with nearly as many unknowns, but with less than 1% of the
coefficients nonzero. In this paper we indicate the optimization criteria used to specify the sought-after solution
and the computational techniques that have been found successful to achieve practically useful solutions at reasonable computer cost.

1. Outline. This is a tutorial paper. It serves as an
introduction to its topic, but it does not attempt to provide a complete survey. Due to the already large literature, such a survey would either be very long, or tantali-
zingly uninformative. References are provided to enable
the reader to carry out an exhaustive follow-up study.
These references, in conjunction with their own references,
cover the work done to date.

In the next section the problem of image reconstruction
from projections is described and the relationship of the
problem to sparse matrices is explained. In Section 3, by
the way of a representative example, a particular approach
(Bayesian optimization) is introduced, and two different
solution strategies (Richardson's method and a row generation method) are explained. In Section 4 we look at a

*Department of Computer Science, State University of New
York at Buffalo, Amherst, New York 14226.

problem of the immediate future: truly three-dimensional
reconstruction from two-dimensional projections. A possible
solution strategy for this "huge" (larger than "large")
problem is discussed. The paper concludes with a brief discussion of the utility in practice of the methods that are
presented in its body.

2. Image Reconstruction from Projections. We use the
word "picture" to describe a function of two variables whose
value is 0 outside a known region (the picture region).
Without loss of generality, we assume that the picture
region is square. The values of a picture represent some
physical parameter; for example, an X-ray attenuation coefficient in a cross-section of the human body or the radio-
brightness in a portion of the sky. Except when referring
to a particular application, we call the value of the picture at a point the density of that point.

In this terminology, a rough statement of the image reconstruction problem is: given estimates of the total
density of a picture in thin strips of known location,
estimate the picture. (See Figure 1.)

This basic problem occurs in a surprisingly large number
of areas in science and medicine. In [1] the reader can
find articles on the reconstruction of the structure of the
solar corona, the radio-brightness of the sky, the distribution of radiation-emitting isotopes inside the human body
and the structure of the dynamic human physiology, including
the beating heart. Figure 2 shows a reconstruction, from
data collected by an electron microscope, of the tail of
a virus.

Probably the most significant application is what has
become known as Computed Tomography (CT). Since there are
a number of introductory articles explaining this area
(e.g., [2]), here we give only a brief overview.

The phrase "tomography" has long been used in medicine
to describe processes which attempt to image a single

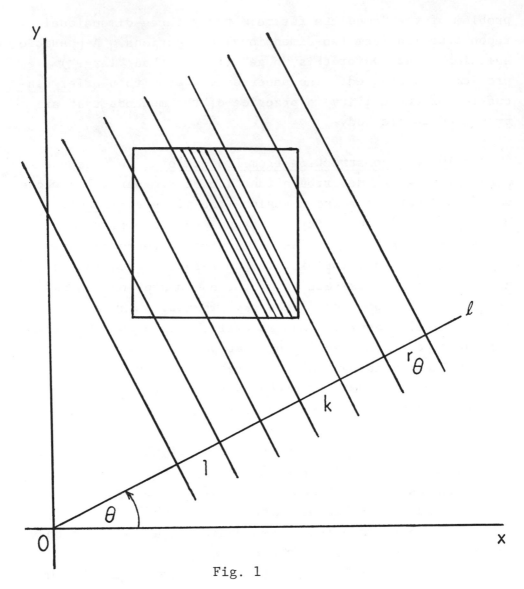

Fig. 1

The input to the image reconstruction algorithm consists
of estimates of the total density of a picture in thin
strips of the type shaded in the figure, for different
values of k and θ. Figure has appeared in: "Two Direct
Methods for Reconstructing Pictures from Their Projections -
A Comparative Study" by G.T. Herman, Computer Graphics and
Image Processing, v. 1, (1972), pp. 123-144.

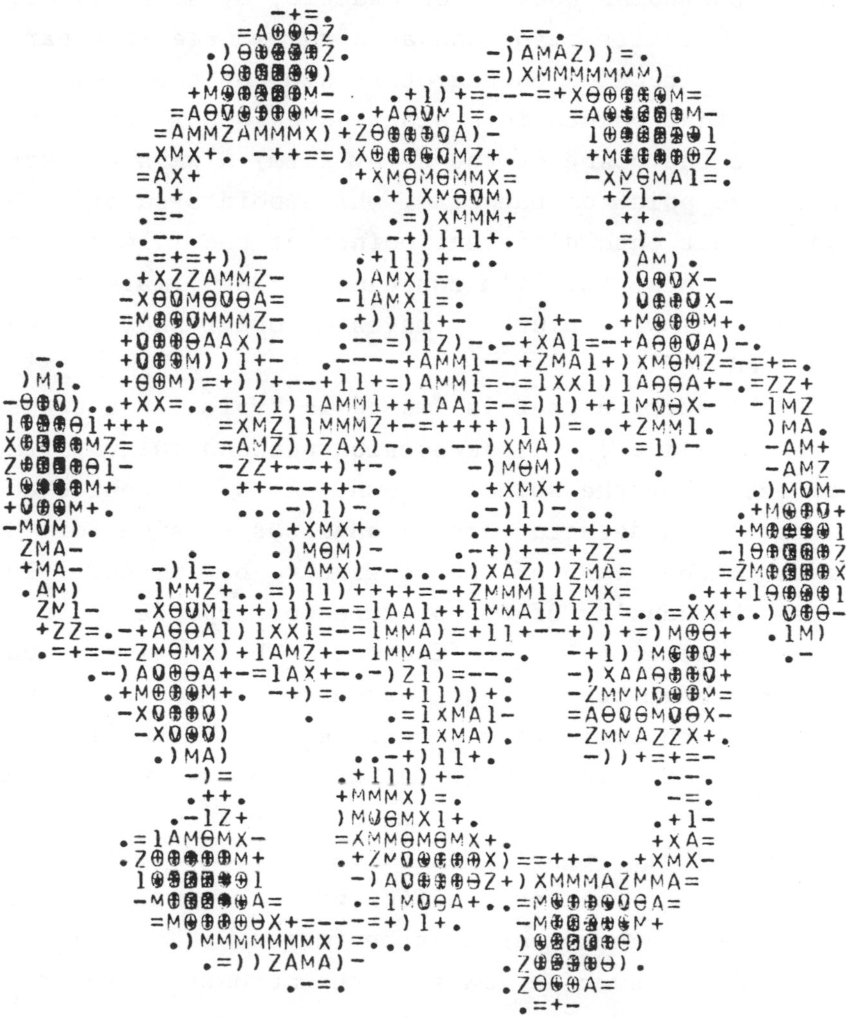

Fig. 2
Cross-section of the tail of the bacteriophage T4.

section of the human body.  For example, by steadily moving
an X-ray film in its plane and an X-ray source in a parallel
but opposing direction on the other side of the body, one
can achieve that points in a single section of the body
(parallel to the plane of the X-ray film) always project
onto the same point on the film, while points outside this
section project onto different points of the film as the
film moves.  Thus, the film contains a sharp image of a
single section, with blurred images of other sections
superimposed on it.

In CT the section that is imaged is typically a trans-
verse section, i.e., a cross-section perpendicular to the
main direction of the spinal column.  An X-ray source and
a detector (or a detector array) on opposite sides of the
body move in the plane of the section to be imaged.   In
this case the density of the picture corresponds to the
X-ray linear attenuation coefficient.  The total attenuation
of the X-ray beam between the source and the detector provides us with an estimate of the integral of the linear
attenuation coefficient in "thin strips of known location".
For details see [3,4].  Since the tissues in different
organs (and in different types of tumors) have different
X-ray  linear attenuation coefficients, the internal structure of the human body can thus be imaged.

An example of such an image is the reconstructed cross-
section of the head of a patient shown in Figure 3.  Typically,  a sequence of such cross-sections is  produced,
giving us the capability to image the three-dimensional
appearance of internal structures of the human body, see
Figures 4 and 5. This combination of CT and computer processing has given us a diagnostic tool of amazing potential,
which has been aptly described by the phrase "noninvasive
vivisection" [5].

There are many different ways of solving the reconstruction problem (see, e.g., [1,4]).  The methods described in
this paper are not necessarily the ones to choose in a par-

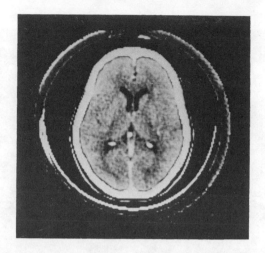

Fig. 3

Reconstructed cross-section of the head of a patient.
Picture supplied by Dr. W.R. Kinkel, Director of Dent
Neurological Institute, Millard Fillmore Hosp., Buffalo, NY.

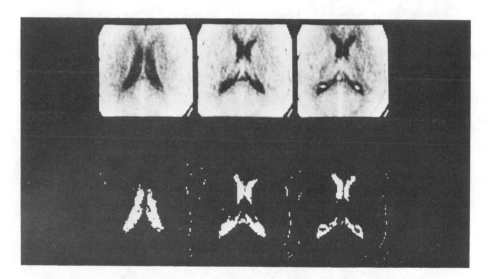

Fig. 4

Parts of the display of cross-sectional slices of the head
of a patient obtained using computerized tomography at the
Dent Neurological Institute of Millard Fillmore Hosp., Buffalo, New York. The ventricles in the brain appear as dark
in the top row and as bright in the bottom row.

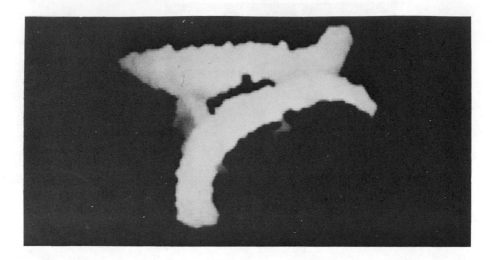

Fig. 5

From displays such as those in Fig. 4, the surface of the
ventricles is computer detected and the three-dimensional
appearance of the ventricles is displayed.  This figure as
well as Fig. 4 have appeared in: "Boundary detection in
3-dimensions with a medical application" by E. Artzy and
G.T. Herman, Technical Report MIPG9, Medical Image Processing Group, Dept. of Computer Science, State University of
New York at Buffalo, Buffalo, N.Y.

---

ticular application area.  In fact, the reconstructions in
Figures 2 and 3 have been obtained using an entirely different method, the so-called convolution method [1,4]. However, in certain cases methods of the kind described below
were found to be the most appropriate, and sometimes the
only ones which were applicable.

The methods we discuss fall into the general class of
reconstruction methods which have been referred to  as
"series expansion methods".  They usually involve an iterative procedure for the solution of a large scale optimization problem.  A precise description of the series expansion methods and a thorough survey (up to early 1976) of
the optimization criteria and iterative approaches that

have been used in image reconstruction can be found in [6].
Here we restrict our attention to a particular approach,
which is a typical example of what is done in this area.
Another approach is discussed in Section 4.

How does the image reconstruction problem translate into
a problem involving sparse matrices? We illustrate this
using the application area of CT.

Consider Figure 6. We subdivide the picture region
into small squares of equal size. These are called pixels,
short for "picture elements". We assume that in any one of
the pixels the density is uniform. The validity of this
assumption depends on the size of the pixels; the smaller
they are, the less our assumption is likely to be violated.
We also assume that our measurements provide us with estimates of line integrals of the density. Then each measurement gives rise to an equality

$$p_i = \sum_{j=1}^{J} r_{i,j} \, x_j + e_j \tag{1}$$

where $p_i$ is the estimated line integral for the ray in
question (based on the measurement), $x_j$ is the assumed
constant density in the j'th of the J pixels (these are
our unknowns), $r_{i,j}$ is the length of the intersection of
the ray with the j'th pixel (these can be calculated from
the geometry), and $e_j$ is the error term, containing both
measurement and discretization error. We use I to denote
the number of measurements, thus $1 \le i \le I$ in Eq. (1). We use
vector notation to rewrite the system

$$p = Rs + e \tag{2}$$

In order to get "good" resolution (and also so that our
assumption of uniformity in the pixels is not too badly
violated) we need a large number of pixels. In CT an
array of $500 \times 500 = 2.5 \times 10^5$ is not untypical. We need a
correspondingly large number of rays. Hence we are dealing
with a very large problem. On the other hand the matrix is
very sparse, since a ray typically intersects approximately

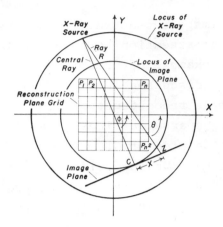

Fig. 6

Diagram of reconstruction plane geometry. Multiple views
of object to be reconstructed can be obtained by rotating
X-ray source and image detector about object.  A plane
transecting irradiated object is mathematically partitioned
into n × n grid of elements, each of which is assumed to
have uniform roentgen opacity.  Grid is constructed so as
to contain entire cross-sectional area of irradiated object.
Measurement of intensity of transmitted ray R  at point Z
and geometric determination of path of ray R (using X,$\phi$,
and $\theta$) define an equation  involving roentgen absorption
values of grid elements which intersect ray path.  Similar
equations for many rays at each position of X-ray source
can be derived and used to resolve unknown roentgen absorption values of each grid element.  Figure as appeared in:
"Three-dimensional visualization of the intact thorax and
contents: A technique for cross-sectional reconstruction
from multiplanar X-ray views" by R.A. Robb, J.F. Greenleaf,
E.L. Ritman, S.A. Johnson, J.D. Sjostrand, G.T. Herman and
E.H. Wood, Computers and Biomedical Research, v. 7 (1974),
pp. 395-419.

$\sqrt{J}$ of the J pixels.

Apart from the size of the problem, we are also faced with the question: "what is to be considered a solution of Eq. (2)?". A survey of answers can be found in [6]; in the next section we look at one of these answers in detail.

3.  The Bayesian Approach to Image Reconstruction. This approach was first proposed by Hurwitz [7]. Our brief outline below is based on [8,9].

We assume that both the image vector x and the error vector e are samples of random variables of known statistical properties. Let $p_X(x)$ and $p_E(e)$ denote the probability density functions of these random variables. For example, we may assume that X is a multivariate Gaussian distribution with mean $\hat{x}$ and E is a multivariate Gaussian distribution with mean the zero vector. For this example,

$$P_X(x) = C \ e^{-\frac{1}{2}(x-\hat{x})\beta^{-1}(x-\hat{x})} \tag{3}$$

and

$$P_E(e) = D \ e^{-e^T Ee} \tag{4}$$

where $\beta$ and E are positive definite matrices and C and D are constants.

The Bayesian approach to image reconstruction selects (given projection data p) the vector x which maximizes

$$P_X(x) \ P_E(p-Rx). \tag{5}$$

If we now assume that $P_X$ and $P_E$ are given by (3) and (4), respectively (the validity of such an assumption is discussed below), then we can prove [8] that the x which minimizes (5) is the same as the x which satisfies

$$(U + 2\beta R^T ER) \ (x-\hat{x}) = 2\beta R^T E(p-R\hat{x}), \tag{6}$$

U is the J×J identity matrix.

Any method for solving the system of equations (6) would now serve in principle. Due to the size of the problem (recall that J is often greater than $10^5$), methods which are computationally feasible need to be carefully selected.

One such method is based on the well-known iterative procedure of Richardson. It can be shown [10] that Richardson's algorithm applied to (6) has the form

$$x^{(0)} = \hat{x},$$
$$x^{(n+1)} = x^{(n)} + \frac{1}{\sigma} (-x^{(n)} + \hat{x} + 2\beta R^T E(p - Rx^{(n)})) \qquad (7)$$

We know [8] that provided $\sigma > (\lambda_{max} \beta R^T ER) + 0.5$, where $\lambda_{max}$ denotes maximum eigenvalue, this algorithm converges to the unique solution $x^*$ of (6); in fact, there is a constant b (which depends only on $\beta$) and a constant $\mu$, such that $0 \leq \mu < 1$ and

$$||x^* - x^{(n)}|| \leq \mu^n b \; ||x^* - \hat{x}||, \qquad (8)$$

where $||\cdot||$ denotes the Euclidean (square) norm.

Furthermore, a single iterative step of the algorithm is, in our application area, relatively simple to implement. This is because it is reasonable to assume that E is a diagonal matrix (i.e., that errors in the measurements are uncorrelated), with i'th diagonal entry $E_{i,i}$. Then

$$R^T E(p - Rx^{(n)}) = \sum_{i=1}^{I} E_{i,i} \; (p_i - R_i^T x^{(n)}) \; R_i, \qquad (9)$$

where $R_i$ is the transpose of the i'th row of the matrix R. Since the components of $R_i$ are the lengths of intersections of the i'th ray with the different pixels, the location and size of the nonzero components of $R_i$ can be calculated fairly easily based on the geometry of the ray and the pixels. For this reason, the matrix R need not be stored at all; which is just as well, since it can be of size over $10^{10}$ with something like $10^8$ nonzero entries. Using the expression in (9), we see that $R^T E(p - Rx^{(n)})$ can be calculated by a single sweep through the rays; i.e., a loop through $1 \leq i \leq I$.

The value of the expression in (9) is a vector of dimensions J, which can be represented as a digitized picture. This vector is multiplied by β. The (u,v)'th entry of β indicates the covariance between the u'th and the v'th pixel of pictures in the assumed distribution. If we assume that $\beta_{u,v} = 0$ if the u'th and v'th pixel are not near each other, multiplication of a vector x by β can be implemented by smoothing the digitized picture represented by x (i.e., replacing each pixel density by a weighted average of itself and its neighbors). Thus, in spite of the size of the vectors and matrices involved, implementation of (7) is relatively simple.

So far, things look good. In fact, there is even a way of selecting σ so that the value of μ in (8) is minimal. These values are [8]

$$\sigma = (\lambda_{max}\beta M^T EM) + (\lambda_{min}\beta M^T EM) + 1 \tag{10}$$

and

$$\mu = \frac{\lambda_{max}\beta M^T EM - \lambda_{min}\beta M^T EM}{\lambda_{max}\beta M^T EM + \lambda_{min}\beta M^T EM + 1}, \tag{11}$$

where $\lambda_{min}$ denotes the smallest eigenvalue. Unfortunately, in practice $\lambda_{max}\beta M^T EM$ is large (e.g., 3000), $\lambda_{min}\beta M^T EM$ is small (e.g., very near 0), hence the calculated "optimal" rate of convergence is slow (e.g., μ in (7) has value 0.9996).

Richardson's method is rarely the method of choice for solving a system of equations. Other iterative procedures such as steepest descent, conjugate gradient, semi-iterative methods and a nonstationary version of Richardson's method have been investigated [11] for their use in solving a system like (6). All these methods can be implemented using the type of procedure we described above, although some of them need to use a larger number of vectors of size J than others, a consideration in view of the size of J ($>10^5$).

While the practical rate of convergence for these methods is different, it was found that the use of "tricks" (rather than of strictly justifiable mathematical procedures) resulted in the greatest improvements. Such tricks include the use of constraining (e.g., setting negative components to 0 at the end of each iteration, knowing that densities are never negative) and selective smoothing (enhancing edges but smoothing the rest of the picture produced at the end of each iteration). It has been found [11] that such tricks work best in combination with Richardson's method with a variable sequence of $\sigma$'s optimized based on the total number of iterations. Even so, since each iterative step is fairly expensive and a fair number of steps may be required, such methods can easily eat up computer time; an hour on a large machine (such as a Cyber 173) can be expected. This rules out their use in medical practice, unless they are implemented on specially programmed microprocessors.

Even more important than the problem of implementation is the question of the clinical efficacy of the image produced by such procedures.

In [8] the Bayesian optimization was applied to a picture representing the thorax of a dog, see Figure 7. It was assumed that in (6) $\hat{x}$ is a uniformly grey picture and that only pixels which touch each other are correlated. Thirteen iterative steps of (7) produced the picture in Figure 8. This compares unfavorably to the picture produced by a much less expensive method (called ART3) shown in Figure 9. Figure 10 gives a quantitative comparison of the reconstructed values produced by the two methods.

Why is the reconstruction produced by the Bayesian method inferior to the other one, when it is supposed to be (an iterative approximation to) the "optimum"? The answer, of course, is that the "optimum" is optimum only according to a mathematical criterion based on certain assumptions about the world which may in fact be false.

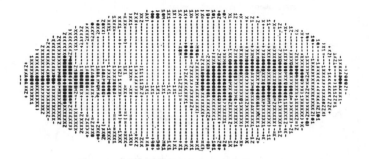

Fig. 7

The thorax of a dog. Half-tone display of original chest phantom produced by a computer printer.

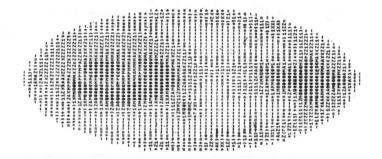

Fig. 8

Reconstruction from 60 views using the Bayesian algorithm.

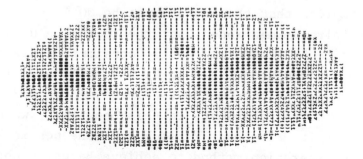

Fig. 9

Reconstruction from 60 views using ART3.

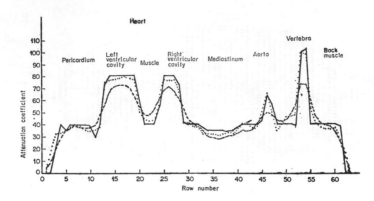

Fig. 10

Graph of exact density values of the 63×63 digitized version of the original shown in Fig. 7 (solid line), reconstruction using the Bayesian algorithm shown in Fig. 8 (broken line) and reconstruction using ART3 shown in Fig. 9 (dotted line) along the 36th row. This figure as well as Fig. 7, 8, and 9 appeared in: "A computer implementation of a Bayesian analysis of image reconstruction" by G.T. Herman and A. Lent, Information and Control, v. 31 (1976), pp. 364-384.

Do $P_X$ and $P_E$ satisfy (3) and (4); and, if they do, are our choices of $\hat{x}$, β and E correct? The inferior result in Figure 8 implies that the answer must be NO.

We have already mentioned that the technique used to produce Figure 9, ART3, is actually considerably less expansive than the method based on Richardson's algorithm. ART3 is a row-generation method. Since such methods are discussed in detail elsewhere in this book [12], we say no more about it. Instead, we look into the possibility of using a row-generation method to achieve Bayesian optimization.

For this discussion we make a further simplification of our assumed model of the world: we assume that both E and β are diagonal matrices with constant entries. (The justi-

fication, or lack of it, for this step is discussed in [9].)
Then (6) can be rewritten as

$$(U + r^2 R^T R)(x - \hat{x}) = r^2 R^T (p - Rx). \tag{12}$$

The following iterative algorithm has been proved [9] to
produce a sequence of $x^{(n)}$'s which converge to a solution
of (12).

$$
\left.
\begin{aligned}
x^{(0)} &= \hat{x} \\
u^{(0)} &= \text{(J-dimensional zero vector)} \\
x^{(n+1)} &= x^{(n)} + r^2 d^{(n)} R_i^T \\
u_i^{(n+1)} &= u_i^{(n)} + d^{(n)} \\
u_k^{(n+1)} &= u_k^{(n)}, \quad \text{if } k \neq i,
\end{aligned}
\right\} \tag{13}
$$

where $i = (n \bmod I) + 1$, and

$$d^{(n)} = p^{(n)} \frac{p_i - \left(u_i^{(n)} + R_i^T x^{(n)}\right)}{1 + r^2 \, ||R_i||^2} \tag{14}$$

with $0 < \underline{\lim} \, p^{(n)} \leq \overline{\lim} \, p^{(n)} < 2$.

Note that this algorithm uses only one row of R in any
iterative step. As discussed above, the location and value
of the nonzero components of $R_i$ can be calculated as and
when they are needed. The algorithm is very efficient on
storage. Only the vector $x^{(n)}$ is such that its components
are randomly accessed, all other items can be kept on a sequentially accessed auxilliary device.

In Figure 11 we show the result of applying the row-
generation algorithm. The top left corner contains a picture representing the human skull and its contents. Below
it is a reconstruction produced by the convolution method
(for details see [9]). Using the result of the convolution
method as $\hat{x}$, we applied 3I iterations of algorithm (13).
The result obtained is in the bottom right corner; the improvement is hardly noticeable. In the top right corner we

show a picture produced by the combination of (13) and "tricks". The "tumor" in the center lower half of the picture is observable in this reconstruction (this is assuming that the photograph reproduced well), while it certainly was not visible in the others.

As far as timing goes, on a sequential computer I iterations of (13) cost about the same as 1 iteration of (7), and in the case of our example (which is fairly typical) we needed only 3I iterations. In the author's experience, the row-generation method appears to be preferable to Richardson's procedure on the basis of computer time, computer storage and quality of the pictures produced.

4. <u>Truly Three-Dimensional Image Reconstruction</u>. Under certain circumstances, for example, if one wants to image the moving heart or lung, it is important to collect the data for a whole three-dimensional distribution spontaneously [5]. One possible arrangement is to have X-ray

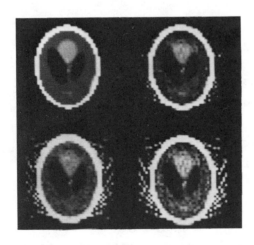

Head phantom and Bayesian reconstructions: 39×39 digitized displays. Top left: phantom; top right: reconstruction by the Bayesian method with tricks; bottom left: reconstruction by a convolution method; bottom right: reconstruction by the Bayesian method with the output of the convolution method as expected

Fig. 11

value $\hat{x}$. This figure appeared in: "On the Bayesian approach to image reconstruction" by G.T. Herman, H. Hurwitz, A. Lent and H.P. Lung, Technical Report 134, Dept. of Computer Science, State University of New York at Buffalo, Buffalo, NY.

sources which project a cone-beam of X-radiation through the body onto a two-dimensional detector. In this section we describe, based on [13], a possible way of reconstructing from such data.

A series expansion approach is used. Let f denote the function of three variables describing the distribution we wish to reconstruct. We take a fixed set of J functions $f_j$ and attempt to approximate f as a linear combination of them.

$$f = \sum_{j=1}^{J} x_j f_j, \tag{15}$$

where $x_j$ are unknowns to be determined. If we had detailed our approach in Section 2 in the same way, then $f_j$ would have been a function which has value 1 in the j'th pixel and value 0 elsewhere.

We number the I rays between the source positions and detector positions by the index i, and use $R_i$ to denote the operation of taking a line integral along the i'th ray. Then the measured integral $p_i$ along the i'th ray can be expressed as

$$p_i \simeq R_i f \simeq \sum_{j=1}^{J} x_j \, R_i f_j = \sum_{j=1}^{J} r_{i,j} \, x_j \tag{16}$$

where $r_{i,j} = R_i f_j$ can be calculated for each ray i and each function $f_j$. In matrix notation we get again (2), with the error vector e taking care of all the approximations in (16).

A criterion to use for selecting a solution (alternative to the Bayesian optimization criterion) is to select a least squares solution; i.e., an x for which

$$||Rx-p||^2 \tag{17}$$

is as small as possible. As is well known, x minimizes (17) if, and only if,

$$R^T Rx = R^T p. \tag{18}$$

In the truly three-dimensional problem we wish to reconstruct a whole three-dimensional distribution. With the
same resolution in cross-sectional slices as we discussed
before and with 40 slices, the number of components of $x$
is of the order of $10^7$ and so $R^T R$ may well be a $10^7 \times 10^7$
matrix. In order to make solution of (18) computationally
feasible, we have to make sure that $R^T R$ has some special
computationally advantageous properties.

Our approach to achieving this is to select both the
$f_j$'s and the positioning of the rays in a clever fashion.
In cylindrical coordinates

$$f_j(\rho,\phi,z) = P_n(\rho)\Phi_m(\phi)Z_k(z), \tag{19}$$

where, for $1 \leq n \leq N$, $P_n(\rho)$ is 1 in a thin cylindrical shell
and 0 otherwise, for $0 \leq m \leq 2M$, $\Phi_m(\phi)$ is harmonic function
of $\phi$ and, for $1 \leq k \leq K$, $Z_k(z)$ is 1 in a thin layer and zero
outside. Note that $I = N(2M+1)K$. The source positions are
chosen to lie equispaced on a circle in the plane $z = 0$.
Under such circumstances (for details, see [13] and its
references) $R^T R$ is a block-diagonal matrix of M+1 blocks.
Each of these blocks itself is a block poly-diagonal, with
K sub-blocks on the diagonal of a block. The number of
nonzero sub-blocks in any row would depend on the number
of slices a ray goes through; it is likely to be small in
practice.

Thus the problem of solving (18) in this case decomposes
into solving M+1 considerably smaller problems, which themselves have a relatively nice structure. In view of this,
solution of the truly three-dimensional problem is not
expected to be essentially more complicated than reconstructing the same body cross-section by cross-section,
from data collected entirely in the cross-sections.

5. Discussion. The methods presented here translate the
problem of image reconstruction from projections into a
sparse matrix problem and then proceed to solve the sparse
matrix problem (in spite of its enormous size) by making

use of special properties of the matrix provided to us by
the geometry and physics of the original situation.

In cases where data have been collected in a uniform
fashion and are plentiful and reliable, alternative reconstruction methods [1,4] produce comparable results at a
fraction of the cost. However, the series expansion approach is applicable to any mode of data collection and
our knowledge of the world can be conveniently incorporated
into an optimization criterion and/or tricks. Thus, the
techniques described in this paper are immediately available
for use for new ways of data collection (other techniques
are often specialized from this point of view), and lack of
quantity or quality in the data can often be compensated for
by introduction of a priori information (which is often not
incorporable into the faster techniques). For a more detailed discussion, see [4,6].

Acknowledgements. The research of the author is supported by NIH grants HL18968, HL4664 and RR7 and NCI grant
CB-84235. He is grateful to his colleagues in the Medical
Image Processing Group, Department of Computer Science,
State University of New York at Buffalo, without whose work
most of what has been described above would not be known
today.

References

[1]    HERMAN, G.T.   (Ed.), Image Reconstruction From
       Projections:   Implementation and Applications,
       Springer-Verlag, Berlin, to appear.

[2]    GORDON, R., JOHNSON, S.A., AND HERMAN, G.T., Image
       reconstruction from projections, Sci. Amer.,
       Oct. (1975), pp. 56-68.

[3]    HERMAN, G.T., An introduction to some basic mathe-
       matical concepts on computed tomography, in:
       Roentgen-Video-Techniques for Dynamic Studies of
       Structure and Function of the Heart and Circu-
       lation (ed. P. Heintzen), Georg Thieme Publishers,
       Stuttgart, Germany, 1978.

[4]    HERMAN, G.T., Image Reconstruction From Projections:
       The Fundamentals of Computerized Tomography,
       Academic Press, New York, to appear.

[5]    WOOD, E.Y., New vistas for the study of structural
       and functional dynamics of the heart, lungs, and
       circulation  by noninvasive numerical tomographic
       vivisection, Circulation 56, (1977), pp. 506-520.

[6]    HERMAN G.T. and LENT, A., Iterative reconstruction
       algorithms, Comp. Biol. and Medicine 6, (1976),
       pp. 273-294.

[7]    HURWITZ, H., Jr., Entropy reduction in Bayesian
       analysis of measurements, Phys. Rev. A, 12,
       (Aug. 1975), pp.  698-706.

[8]    HERMAN, G.T. and LENT, A., A computer implementation
       of a Bayesian analysis of image reconstruction,
       Info. and Control 31, (1976), pp. 364-384.

[9]    HERMAN, G.T., HURWITZ, H., LENT, A. and LUNG, H.P.,
       On the Bayesian approach to image reconstruction,
       Technical Report 134, Dept. of Computer Science,
       S.U.N.Y. at Buffalo, Amherst, N.Y., 1978.

[10]   HERMAN, G.T. and LENT, A., Quadratic optimization
       for image reconstruction, Part I, Comp. Grph. and
       Imge. Proc. 5, (1976), pp. 319-332.

[11]   ARTZY, E. and HERMAN, G.T., Investigation of
       quadratic optimization techniques for image recon-
       struction, Proceedings of the 1977 IEEE Conference
       on Decision and Control, New Orleans, La. (Dec. 1977),
       pp. 350-360.

[12]   CENSOR, Y. and HERMAN, G.T., Row-generation methods
       for feasibility and optimization involving sparse
       matrices and their applications, These proceedings.
       Also Technical Report MIPG18, Medical Image Proces-
       sing Group, SUNY at Buffalo, Amherst, N.Y., 1978.

[13]   ALTSCHULER, M., HERMAN, G.T.,  and LENT, A.,
       Fully three-dimensional image reconstruction from
       cone-beam sources, Proceedings of the IEEE Computer
       Soc. Conf. on Pattern Recognition and Image Proc.,
       (1978), pp. 194-199.

# Row-Generation Methods
# for Feasibility and Optimization Problems
# Involving Sparse Matrices
# and Their Applications

## Yair Censor* and Gabor T. Herman**

Abstract. This paper brings together and discusses
theory and applications of row-generation methods for linear
feasibility problems (find $x \in R^n$, such that $Ax \leq b$), linear
inequalities constrained optimization problems (minimize
$f(x)$, subject to $Ax \leq b$) and some interval convex programming problems (minimize $f(x)$, subject to $c \leq Ax \leq b$).
A row-generation method is any iterative procedure which,
without making any changes to the original matrix A, uses
the rows of A, only one row at a time. Such methods are
important and have demonstrated their effectiveness in
problems with large or huge matrices which do not enjoy any
detectable structure pattern apart from a high degree of
sparseness.
Fields of application where row-generation methods are
used in various ways include Image Reconstruction from
Projections, Operations Research and Game Theory, Learning
Theory and Pattern Recognition, and Transportation Theory.

1. Introduction. In some fields of applications the

modelling of the physical problem leads to a system of

linear equations

(1.1)     $\langle a_i, x \rangle = b_i$ ,    $i \in I$ $\left( I \equiv \{1, 2, \ldots, m\} \right)$,

where $a_i$ and x belong to the n-dimensional Euclidean space

$R^n$ and $\langle \cdot, \cdot \rangle$ stands for the usual inner product. This

system is often inconsistent, or we have reason to believe

that its exact algebraic solution is less desirable, in

terms of the physical problem, than some other, differently

*Medical Image Processing Group, Dept. of Computer Science,
 State University of New York at Buffalo, Amherst, NY 14226.
**Dept. of Computer Science, State University of New York
 at Buffalo, Amherst, NY 14226.

defined "solution".

Such a belief may be nourished by evidence about measurements inaccuracy, noise corruption of data, discretization in the model, etc. In such cases it is sometimes useful to seek a point that will lie in a specified vicinity of all hyperplanes defined by the equations of (1.1). Two approaches may be taken: (i) The Feasibility Approach.

(1.2)   Find $x \in R^n$ such that $b - \delta \leq Ax \leq b + \varepsilon$.

Here A is the m×n matrix having $a_i^T$ in its i-th row, $b = \begin{pmatrix} b_1 \\ \vdots \\ b_m \end{pmatrix}$,

and $\delta$ and $\varepsilon$ are some prescribed tolerance vectors.

(ii) The Optimization Approach.

(1.3)   Find $x \in R^n$ which minimizes $f(x)$ and is such that

$b - \delta \leq Ax \leq b + \varepsilon$.

Here $f(x)$ is some predesignated objective function. Both (1.2) and (1.3) are interval constrained problems. When no advantage can be taken of the fact that the inequalities come in pairs they can be viewed as linearly constrained problems with a (twice as large!) system of (one-sided) linear inequalities.

2. The Special Environment and The Role of Row-Generation Methods. The environment within which these problems are treated is distinguished by a combination of properties. (i) Dimensionality. The system (1.1) is huge, e.g, A is m×n with $n \geq 10^5$ and m even greater. (ii) Sparseness. A is sparse, e.g., less than 1% of the entries are nonzero. (iii) Lack of Structure. We fail to recognize any

structure pattern in the distribution of nonzero entries throughout the matrix. Alternatively, sometimes we might detect a structure, but either we have no idea how, or we cannot afford to take advantage of it. (iv) Time Restriction. In some applications (image reconstruction from projections for medical diagnosis is an example) there might be an inherent time restriction, i.e., a solution which is acceptable in terms of the physical problem is demanded within minutes of data collection.

In such an environment the use of row-generation methods, rather than other methods such as are described in [17,18,41,45], is strongly suggested, because their properties enable them to retain the sparseness, cope with dimensionality of the problem, and take advantage of the sparsity without relying on any particular structure pattern. Experience in different fields shows their power to reach acceptable solutions quickly [27,30].

(2.1) Definition. A row-generation method (RG method, for short) is an iterative algorithm which has the following properties: (i) no changes are made to the original matrix; (ii) no operations are performed on the matrix as a whole; (iii) in a single iterative step access is required to only one row of the matrix; (iv) the algorithm presents modest arithmetical demands.

Several remarks concerning this definition seem to be in order. It is sometimes possible to avoid storing the matrix explicitly at all and have the i-th row generated from the experimental data each time anew. We refer to this as row-generation capability. (iii) makes RG methods favorable in such cases. Without such capability one would, of course, consult any of the efficient methods for storing a sparse matrix (see, e.g., [13]). In keeping the vague condition (iv) in Definition (2.1) we try to make the point that in our special environment (particularly dimensionality), performing operations which are just a little more complicated might quickly reduce the practicality of a method.

In view of the fact that the matrix is not changed and not operated on as a whole, and that sometimes it is not even stored any more we like to make

(2.2) A Teasing Claim. Can it be that beyond some dimension barrier there is no matrix at all any more?

We mean that while the system (1.1) is always there, it seems that for practical implementation the matrix ceases to play any useful role.

3. Some Common Notions.

(3.1) Control of an RG method. This is a sequence of indices $\{i_k\}_{k=0}^{\infty}$ according to which the rows of the matrix A are taken up, i.e., in the step $k \rightarrow k+1$ the $i_k$-th row is used.

(3.2) Examples (see, e.g., [24]). (i) Cyclic Control: $i_k = k \pmod{m} + 1$ where m is the number of rows in A.

(ii) Control of the Remotest Set: this is obtained by taking $d(x^{(k)}, H_{i_k}) = \underset{i \in I}{\text{Max}}\, d(x^{(k)}, H_i)$ where $d(x^{(k)}, H_i)$

represents the Euclidean distance from the iterate $x^{(k)}$ to the hyperplane $H_i = \{x | <a_i, x> = b_i\}$. (iii) <u>Almost Cyclic Control</u> ([42]): $\{i_k\}_{k=0}^{\infty}$ is almost cyclic on $I = \{1, 2, \ldots, m\}$, if $i_k \in I$ for every $k \geq 0$ and there exists an integer $C$ such that for all $k \geq 0$, $i \subseteq \{i_{k+1}, i_{k+2}, \ldots, i_{k+C}\}$. The cyclic control is almost cyclic with $C = m$.

(3.3) <u>Relaxation parameters</u>. These form a sequence of real numbers $\{\lambda_k\}_{k=0}^{\infty}$ (typically $\lambda_k \in (0, 2)$, for all k) and allow us to overdo or underdo (in some sense) the typical step of an RG method.

4. <u>Row-Generation Methods</u>. We give a brief review of some RG methods. For details, convergence, and other results the reader is referred to the literature. The information about each method here is organized according to the following order: (1) the problem it is designed to solve, (2) formulation of the algorithm, (3) a geometric interpretation, and (4) comments and references.

(4.1) <u>The Method of Kaczmarz</u>. <u>Problem</u>: Solve Ax=b, assuming there exists a solution. <u>Method</u>:

Initialization: $x^{(0)}$ arbitrary

Typical Step: $x^{(k+1)} = x^{(k)} + \lambda_k \dfrac{b_{i_k} - <a_{i_k}, x^{(k)}>}{||a_{i_k}||^2} a_{i_k}$

Control: almost cyclic

Relaxation: $\varepsilon_1 \leq \lambda_k \leq 2 - \varepsilon_2$, $\varepsilon_1$, $\varepsilon_2 > 0$

<u>Geometric interpretation</u>: Given $x^{(k)}$ and the hyperplane $H_{i_k}$ determined by the $i_k$-th equation, $x^{(k+1)}$ lies on the line through $x^{(k)}$ prependicular to $H_{i_k}$, $\lambda_k$ determines its

exact position.  For unity relaxation, $\lambda_k=1$ for all $k\geq0$, we have

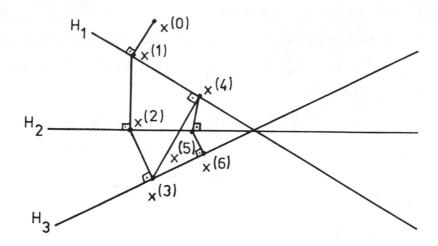

<u>Comments and References</u>:  The method was originally devised by Kaczmarz [39] who proved convergence with unity  relaxation for square nonsingular A.  It was rediscovered by Gordon, Bender and Herman [22] and called ART in the field of image reconstruction from projections [28].  Tanabe studied the method for general matrices [55], its relation to SOR methods was recently pointed out by Bjorck and Elfving [2], and McCormick [46] extended it to nonlinear equations.  See also [3,16,56].

(4.2)  <u>The Method of Successive Orthogonal Projections</u>. This method of Gubin, Polyak and Raik [24] is not an RG method in the sense of Definition (2.1).  Nevertheless both the method of Kaczmarz and the Relaxation Method of Agmon, Motzkin and Schoenberg (4.3, below) are special cases of it.  <u>Problem</u>:  Given a family  $\{Q_i\}_{i\in I}$ of closed and convex sets in $R^n$ with $\bigcap_{i \in I} Q_i \neq \emptyset$ , find a point $x \in \bigcap_{i \in I} Q_i$.

Method:

Initialization: $x^{(0)}$ arbitrary

Typical Step: $x^{(k+1)} = x^{(k)} + \lambda_k \left( P_{Q_{i_k}} (x^{(k)}) - x^{(k)} \right)$

Control: almost cyclic

Relaxation: $\varepsilon_1 \leq \lambda_k \leq 2 - \varepsilon_2$, $\varepsilon_1, \varepsilon_2 > 0$

where $P_Q(x)$ is the orthogonal projection of x onto a set Q.

Remark: Here and in several of the other methods described in this paper, other controls, like the control of the remotest set, are applicable. In compliance with (iv) of Definition (2.1) we keep emphasizing the cyclic and almost cyclic controls.

Geometric interpretation: The figure demonstrates the behavior of the method for unity relaxation.

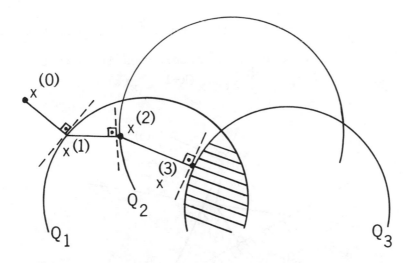

Comments and References: Choosing $Q_i$ to be the hyperplanes of Ax=b we get from this method, the method of Kaczmarz.

The method was generalized (for unity relaxation only) to non-orthogonal projections by Bregman [5], see also [8].

(4.3) <u>The Relaxation Method of Agmon, Motzkin and Schoenberg</u>.

<u>Problem</u>: Find an $x \in R^n$ such that $Ax \leq b$ (linear feasibility problem). <u>Method</u>:

Initialization: $x^{(0)}$ arbitrary

Typical Step: $x^{(k+1)} = x^{(k)} + c^{(k)} a_{i_k}$, where

$$c^{(k)} = \text{Min}(0, \lambda_k \frac{b_{i_k} - \langle a_{i_k}, x^{(k)} \rangle}{||a_{i_k}||^2})$$

Control: almost cyclic

Relaxation: $\varepsilon_1 \leq \lambda_k \leq 2 - \varepsilon_2$, $\varepsilon_1, \varepsilon_2 > 0$

<u>Geometric interpretation</u>: Given $x^{(k)}$ and the closed half-space $Q_{i_k}$ determined by the $i_k$-th inequality, if $x^{(k)} \notin Q_{i_k}$ then $x^{(k+1)}$ lies on the line through $x^{(k)}$ perpendicular to the bounding hyperplane $H_{i_k}$, $\lambda_k$ determines its exact position; but if $x^{(k)} \in Q_{i_k}$, then $x^{(k+1)} = x^{(k)}$.

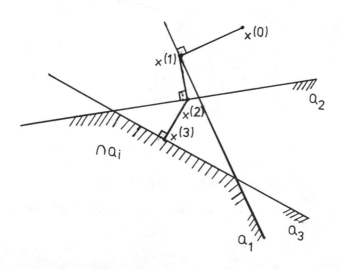

Comments and References: Agmon [1] and Motzkin and Schoenberg [49] published the method in 1954. It is a special case of the method of successive orthogonal projections [24]. If judged on its mathematical contents only, the perceptron convergence theorem seems to be a rediscovery of the method; see p. 248 of Minsky and Papert [48], see also Nilsson [50]. The method was extensively studied by Goffin [19,20,21], who also established its relations with subgradient optimization; see also Polyak [54]. Oettli [52] furnished an appropriate extension of the method to the convex feasibility problem (find $x \in R^n$ such that $f_i(x) \leq 0$ for all $i \in I = \{1,2,\ldots,m\}$). However, an extension of the relaxation method of Agmon, Motzkin and Schoenberg to the convex feasibility problem under cyclic control was obtained only recently, Censor and Lent [9].

(4.4) Hildreth's Algorithm. Problem: Min $\frac{1}{2}||x||^2$ s.t. $Ax \leq b$. Method: (extended version [43]):

Initialization: 
$$\begin{cases} z^{(0)} \text{ is an arbitrary vector in the} \\ \text{nonnegative orthant of } R^m \\ x^{(0)} = -A^T z^{(0)} \end{cases}$$

Typical Step: 
$$\begin{cases} x^{(k+1)} = x^{(k)} + c^{(k)} a_{i_k}, \\ z^{(k+1)} = z^{(k)} - c^{(k)} e_{i_k} \end{cases} \quad \text{where}$$

where $e_{i_k} \in R^m$ and has all components zero except for the $i_k$-th component which is one, and

$$c^{(k)} = \text{Min}\left(z_{i_k}^{(k)}, \ \lambda_k \frac{b_{i_k} - \langle a_{i_k}, x^{(k)} \rangle}{||a_{i_k}||^2}\right)$$

Control: almost cyclic

Relaxation: $\varepsilon_1 \leq \lambda_k \leq 2-\varepsilon_2$, $\varepsilon_1, \varepsilon_2 > 0$

Geometric interpretation: Same as for the method of Agmon, Motzkin and Schoenberg (Section 4.3 above), except that if $x^{(k)} \in \text{int } Q_{i_k}$ then here a move perpendicular to $H_{i_k}$ is made.

Comments and References: Hildreth's method is a primal-dual optimization method; $\{z^{(k)}\}$ is the sequence of dual iterates. [43] contains an extensive list of references related to this method. Hildreth [37] published the method in 1957, see also [11, 25, 44]. Applications were made in [23,57] and some results on rate of convergence were given in [51].

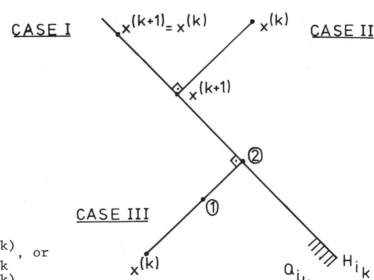

CASE I

CASE II

CASE III

IN CASE III:

① $c^{(k)} = z_{i_k}^{(k)}$, or

② $c^{(k)} < z_{i_k}^{(k)}$

(4.5) <u>Hildreth's Method for Interval Constraints</u>. <u>Problem</u>:

Min $\frac{1}{2}||y||^2$  s.t.  $\gamma \leq Ay \leq \delta$.  <u>Method</u>:

Initialization:
$$
\begin{cases}
y^{(0)} = 0 \\
u^{(0)} = 0
\end{cases}
$$

Typical Step:
$$
\begin{cases}
y^{(k+1)} = y^{(k)} + d^{(k)} a_{i_k} \\
u^{(k+1)} = u^{(k)} - d^{(k)} e_{i_k}
\end{cases}
$$

where $d^{(k)} = \text{Mid} \left( u_{i_k}^{(k)}, \dfrac{\delta_{i_k} - \langle a_{i_k}, y^{(k)} \rangle}{||a_{i_k}||^2}, \dfrac{\gamma_{i_k} - \langle a_{i_k}, y^{(k)} \rangle}{||a_{i_k}||^2}, \right)$

$e_{i_k}$  has one as the $i_k$-th component and zeros elsewhere,

and Mid $(a,b,c)$ stands for the median of the three real

numbers  $a$, $b$, $c$.

Control:  almost cyclic

Relaxation:  always unity, i.e.,  $\lambda_k = 1$ for all $k$

<u>Geometric interpretation</u>: In a typical step the $i_k$-th

interval is employed, i.e., a hyperslab.  Given $y^{(k)}$, the

next iterate will be inside the $i_k$-th slab or on one of its

bounding hyperplanes according to the value of $d^{(k)}$.

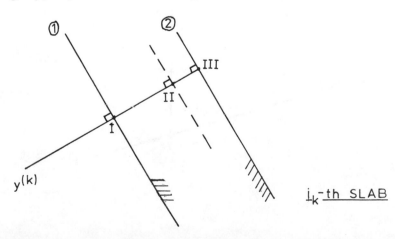

$i_k$-th SLAB

Comments and References: The method was introduced by Herman and Lent [35] and studied further by them in [34]. Its advantages over direct application of Hildreth's method which ignores the interval structure of the constraints are: (i) storage saving because the dual vector is of length m and not 2m; (ii) time saving because half as much iterative steps are needed to get the same approximation.

(4.6) Bregman's Row-Generation Method. Given a function $f:R^n \to R$, Bregman [5] constructs from it another function $D:R^n \times R^n \to R$ in a particular manner. A D-projection of a point $x \in R^n$ onto a set $Q \subseteq R^n$ is then defined as the point y in the set Q which minimizes the value of $D(x,z)$ over all $z \in Q$. For the special choice $f(x) = \frac{1}{2}||x||^2$, the D-function is $D(x,z) = ||x-z||^2$, so that the D-projection in this case coincides with the usual orthogonal projection. In general, other choices for $f(x)$, that will give rise to D-projections which are non-orthogonal, are possible, see Censor and Lent [8].

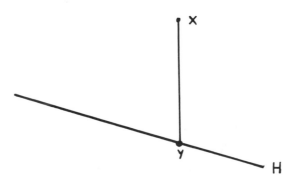

<u>Problem</u>: Min $f(x)$ s.t. $Ax \leq b$. <u>Method</u>:

Initialization: $\begin{cases} x^{(0)} \text{ arbitrary element of a set } Z_0 \\ \text{ defined in [8],} \\ z^{(0)} \text{ is such that } \nabla f(x^{(0)}) = -A^T z^{(0)} \end{cases}$

Typical Step: $\begin{cases} \nabla f(x^{(k+1)}) = \nabla f(x^{(k)}, + c^{(k)} a_{i_k} \\ z^{(k+1)} = z^{(k)} - c^{(k)} e_{i_k} \end{cases}$

and $c^{(k)} = \text{Min}(z_{i_k}^{(k)}, B^{(k)})$ where $B^{(k)}$ is the parame-

ter associated with D-projecting $x^{(k)}$ onto $H_{i_k}$ (see

[8]).

Control: almost cyclic

Relaxation: always unity

<u>Geometric interpretation</u>: Same as that of Hildreth's

method but with D-projections instead of orthogonal pro-

jections.

<u>Comments and References</u>: For $f(x) = \frac{1}{2}||x||^2$ Bregman's algo-

rithm reduces to Hildreth's method. Bregman [5] appeared

in 1967 and was further studied in Censor and Lent [8]

where several open questions are described. For a general

$f(x)$ the typical step involves a separate "inner-loop"

solution to the nonlinear equation

$\nabla f(x^{(k+1)}) = \nabla f(x^{(k)}) + c^{(k)} a_{i_k}$. The method is appli-

cable to the "xlogx" entropy functional, thereby allowing

the iterative solution of entropy optimization problems of

the form $\begin{cases} \text{Min} \sum_{j=1}^{n} x_j \ell n x_j \\ \text{s.t.} \quad Ax \leq b \end{cases}$

through the successive application of "entropy-projections"
(see [7]).

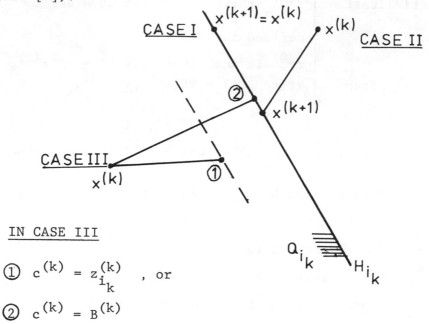

IN CASE III

① $c^{(k)} = z_{i_k}^{(k)}$ , or

② $c^{(k)} = B^{(k)}$

(4.7)  A Row-Generation Method for Interval Convex

Programming.  The idea embodied in the method of

Section 4.5 was recently extended to Bregman's algorithm

by Censor and Lent [8], giving rise to the following.

Problem:

Min $f(y)$  s.t.  $\gamma \leq \Phi y \leq \delta$  Method:

Initialization: $\begin{cases} y^{(0)} \text{ is an arbitrary element of a set } U \\ \text{defined in } [8] \\ u^{(0)} \text{ is such that } \nabla f(y^{(0)}) = -\Phi^T u^{(0)}. \end{cases}$

Typical Step: $\begin{bmatrix} \nabla f(y^{(k+1)}) = \nabla f(y^{(k)}) + d^{(k)} \phi_{j_k} \\ u^{(k+1)} = u^{(k)} - d^{(k)} e_{j_k} \end{bmatrix}$

with $d^{(k)} = \text{Mid } (u_{j_k}^{(k)}, \Delta^{(k)},\ ^{(k)})$ (see [8] for details).

Control:  almost cyclic

Relaxation:  always unity

Geometric interpretation:  Same as in Section 4.5 with D-projections instead of orthogonal projections.

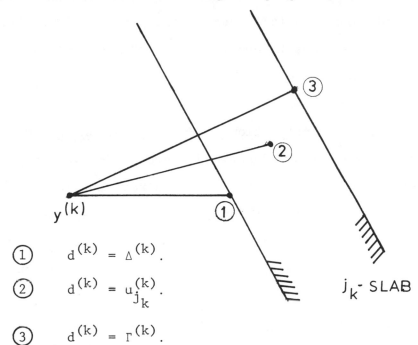

$y^{(k)}$

$j_k$- SLAB

①     $d^{(k)} = \Delta^{(k)}$.

②     $d^{(k)} = u_{j_k}^{(k)}$.

③     $d^{(k)} = \Gamma^{(k)}$.

(4.8) <u>MART</u>:  Another RG Method for Entropy Optimization.

<u>Problem</u>:     $\text{Min} \sum\limits_{j=1}^{n} x_j \ell n x_j$ s.t. $Ax = d$ and $x \geq 0$.    <u>Method</u>:

Initialization:   $x^{(0)} = (1,1,\ldots,1)$

Typical Step:

$$x_j^{(k+1)} = \left( \frac{d_{i_k}}{\langle a_{i_k}, x^{(k)} \rangle} \right)^{\lambda_k a_{i_k,j}} \cdot x_j^{(k)}$$

Control:  Almost cyclic

Relaxation:   $\varepsilon_1 \leq \lambda_k \leq 1, \quad \varepsilon_1 > 0$

Comments and References: MART (≡Multiplicative Algebraic Reconstruction Technique) was suggested by Gordon, Bender and Herman in [22]. Lent [42] supplied a proof of convergence and Elfving studied rate of convergence [14]. Minerbo and Sanderson [47] report on experimental results. A version of MART was in use in transportation theory for a long time, see [2, 40].

## 5. Fields of Application of Row-Generation Methods

Some of the methods described above and/or variants of them were used, either in the way of direct application or by underlying a specific theory, in many diverse fields. In (a) and (b), below, the environment is closest to the one we described in Section 1, and the results there justify our conclusion about the importance of RG methods in such circumstances. (a) Image Reconstruction from Projections. The approach leading to linear feasibility or to linearly constrained optimization in this area is adequately described, along with a survey of algorithms in [29]. The practical problem of computerized X-ray tomography of the human body leads to the very demanding restriction that the amount of time required to obtain an acceptable approximation to the solution should be brief. The matrices here are always huge, sparse and structureless; nevertheless various row generation methods have been shown to be able to handle them successfully [22,30, 31,32,33,34]. See also [36].

(b)  Operation Research. In the well-known travelling salesman problem [26] the subgradient optimization method was introduced to obtain a lower bound on the cost of an optimum tour, from which optimum solutions to the travelling salesman problem can be easily derived.  In this application area the subgradient optimization method can be implemented as a row-generation method.  Large assignment problems and multicommodity network problems were solved in a similar way [6,27]. (c)  Entropy Optimization Problems.  Linearly constrained entropy optimization problems may arise when some model is described in a statistical environment and a solution which is the most objective or maximally uncommitted with respect to missing information is sought for.  Such is the situation in transportation  problems where several row generation methods are used [4,53]. Entropy optimization also appears  in image reconstruction from projections where it is also handled by row generation methods [7,22,42]. Closely related to entropy optimization are linear programming problems augmented by an entropy constraint [15].  (d) Learning Theory. In this discipline, row generation methods underlie the theory describing various modes of behavior of perceptrons [48] and many row generation methods come  into play, as, for example in [38].  See also [50].  (e)  Pattern Recognition.  Relaxation procedures have been found useful in minimizing a criterion function in pattern recognition. For details see [12] and its references.  (f) Game Theory.

Recently [58], the relaxation method was put to work to produce a dynamic theory which describes a bargaining scheme leading to a core solution of an n-person game in characteristic function form. In this way the idea of reaching a core solution of the game by the players having to actually solve the defining system of linear inequalities is replaced by what seems to be a more realistic view, a negotiation process which yields a sequence of payoffs which converge to a payoff in the core of the game.

6.  Row-Generation Methods - A Temporary Conclusion.

Row generation methods for linear feasibility problems and for linearly constrained optimization problems are strongly suggested in an environment of huge, sparse and unstructured matrices for speedy results of acceptable quality. Some RG-methods were successfully applied in various fields and more effort needs to be invested in the study of various RG-methods.

Here we have brought together some important, but by no means all, row-generation methods. We believe that by bringing row-generation methods under one roof, cross-fertilization between various fields of applications where they are used might occur and a better insight into their mathematical nature might be gained [10].

Acknowledgements. We wish to thank our colleague Dr. Arnold Lent who took part in our joint studies upon which this paper is based.  We thank Ms. Barbara S. Dane for typing the manuscript and preparing the drawings.

We acknowledge support from NIH grants HL18968, HL4664, and RR7, and NCI contract CB-84235.

## References

[1] AGMON, S., The relaxation method for linear inequalities, Canad. J. Math., 6 (1954), pp. 382-392.

[2] BJORCK, A. and ELFVING, T., Accelerated projection methods for computing pseudoinverse solutions of systems of linear equations, Technical Report LITH-MAT-R-1977, Dept. of Math., Linkoping Univ., Linkoping, Sweden, (1977).

[3] BODEWIG, E., Matrix Calculus, North-Holland, Amsterdam, 1959.

[4] BREGMAN, L.M., Proof of the convergence of Sheleikhovskii's method for a problem with transportation constraints, USSR Comp. Math. and Math. Phys. 7 (1) (1967), pp.  191-204.

[5] BREGMAN, L.M., The relaxation method of finding the common point of convex sets and its application to the solution of problems in convex programming, USSR Comp. Math. and Math. Phys. 7 (3) (1967), pp. 200-217.

[6] CAMERINI, P.M., FRATTA, L., and MAFFIOLI, F., On improving relaxation methods by  modified gradient techniques.  Math. Prog. Study 3 (1975), pp.  26-34.

[7] CENSOR, Y., LAKSHMINARAYANAN, A.V. and LENT, A., Relaxational methods for large-scale entropy optimization problems with application in image reconstruction, proceedings of the First Annual Workshop on the Information Linkage Between Applied Mathematics and Industry, Naval Postgraduate School, Monterey, Calif., P.C.C. Wang (Ed.), Academic Press (1978).

[8] CENSOR, Y. and LENT, A., An Iterative row-generation method for interval convex programming, Technical Report MIPG11, Medical Image Processing Group, Dept. of Computer Science, SUNY at Buffalo, Amherst, N.Y., Oct. 1978.

[9] CENSOR, Y. and LENT, A., A cyclic subgradient projections method for finding a common minimizer of a family of convex functions over a constraint set and its application, Technical Report MIPG12, Medical Image Processing Group, Dept. of Computer Science, SUNY at Buffalo, Amherst, NY., Sept. 1978.

[10] CENSOR, Y., Row-action methods for huge and sparse systems and their applications, In preparation for SIAM Review.

[11] D'ESOPO, D.A., A convex programming procedure, Nav. Res. Log. Quart. 6 (1959), pp. 33-42.

[12] DUDA, R.O. and HART, P.E., Pattern Classification and Scene Analysis, Wiley, New York, 1973.

[13] DUFF, I.S., A survey of sparse matrix research, Proc. IEEE, 65 (1977), pp. 500-535.

[14] ELFVING, T., A method for computing the maximum entropy solution of a linear system, Technical Report LiTH-MAT-R-1978-4, Dept. of Math., Linkoping Univ., Linkoping, Sweden, (1978).

[15] ERLANDER, S. ,Entropy in linear programs - an approach to planning, Technical Report LiTH-MAT-R-77-3, Dept. of Math., Linkoping Univ., Linkoping, Sweden, (1977).

[16] GASTINEL, N., Linear Numerical Analysis, Hermann, Paris, 1970.

[17] GEOFFRION, A.M., Elements of large-scale mathematical programming, Man. Sci. 16 (1970), pp. 652-691.

[18] GILL, P.E. and MURRAY, W., Methods for Large-Scale Linearly Constrained Problems, Chap. IV in: Gill, P.E. and Murray, W. (Editors), Numerical Methods for Constrained Optimization, Academic Press, New York, 1974.

[19] GOFFIN, J.L., The relaxation method for solving systems of linear inequalities, Working paper no. 77-54, Faculty of Management, McGill Univ., Montreal, Canada, Sept., 1977. (Revised Aug., 1978)

[20] GOFFIN, J. L., On Convergence rates of subgradient optimization methods, Math. Prog. 13 (1977), pp. 329-347.

[21] GOFFIN, J. L., Non-differentiable optimization and the relaxation method, in: Lemarechal, C. and Mifflin, R. (Editors): Nonsmooth Optimization, Pergamon Press, 1978, pp. 31-49.

[22] GORDON, R., BENDER, R. and HERMAN, G.T., Algebraic reconstruction techniques (ART) for three-dimensional electron microscopy and x-ray photography, J. Theor. Biol. 29 (1970), pp. 471-481.

[23] GORENFLO, R., and KOVETZ, Y., Solution of an Abel-type integral equation in the presence of noise by quadratic programming, Numer. Math. 8 (1966), pp. 392-406.

[24] GUBIN, L.G., POLYAK, B.T., and RAIK, E.V., The method of projections for finding the common point of convex sets. USSR Comp. Math. and Math. Phys. 7 (1967), pp. 1-24.

[25] HADLEY, G., Nonlinear and Dynamic Programming, Addison-Wesley, Reading, 1964.

[26] HELD, M. and KARP, R.M., The travelling-salesman problem and minimum spanning trees: part II, Math. Prog. 1 (1971), pp. 6-25.

[27] HELD, M., WOLFE, P. and CROWDER, H.P., Validation of subgradient optimization, Math. Prog. 6 (1974), pp. 62-68.

[28] HERMAN, G.T., LENT, A., and ROWLAND, S.W., ART: mathematics and applications, J. Theo. Biol. 42 (1973), pp. 1-32.

[29] HERMAN, G.T. and LENT, A., Iterative reconstruction algorithms, Compt. Biol. Med. 6 (1976), pp. 273-294.

[30] HERMAN, G.T., LENT, A. and LUTZ, P.H., Iterative relaxation methods for image reconstruction, Comm. Assoc. Comp. Mach.21 (1978), pp. 152-158.

[31] HERMAN, G.T. and LENT, A., A computer implementation of a Bayesian analysis of image reconstruction, Info. and Control 31 (1976), pp. 364-384.

[32] HERMAN G.T., HURWITZ, H. and LENT, A., A Bayesian analysis of image reconstruction, TER-POGOSSIAN, M.M. et al. (Editors), Reconstruction Tomography in Diagnostic Radiology and Nuclear Medicine, University Park Press, Baltimore, Md. (1977), pp. 85-103.

[33] HERMAN, G.T., HURWITZ, H., LENT, A., and LUNG, H.P., On the Bayesian approach to image reconstruction, Technical Report 134, Dept. of Computer Science, SUNY at Buffalo, Amherst, N.Y. (1978).

[34] HERMAN, G.T. and LENT, A., A family of iterative quadratic optimization algorithms for pairs of inequalities with application in diagnostic radiology, Math. Prog. Study 9 (1978), pp. 15-29.

[35] HERMAN, G.T., and LENT, A., A relaxation method with applications in diagnostic radiology, In Survey of Mathematical Programming, A. Prekopa (Editor), North Holland, 1978. (three volumes)

[36] HERMAN, G.T., The use of sparse matrices for image reconstruction from projections, These proceedings. Also Technical Report MIPG20, Medical Image Processing Group, Dept. of Computer Science, SUNY at Buffalo, Amherst, N.Y., Nov. 1978.

[37] HILDRETH, C., A quadratic programming procedure, Naval Res. Log. Quart. 4, (1957), pp. 79-85, Erratum, ibid. p. 361.

[38] JAKUBOVIC, V.A., Finitely convergent recursive algorithms for the solution of systems of inequalities, Doklady, Tom 166 (1966), pp. 300-304.

[39] KACZMARZ, S., Angenaherte Auflosung von Systemen Linearer Gleichungen, Bull. Acad. Polon. Sci. Lett. A, 35 (1937), pp. 355-357.

[40] KRUITHOF, J. Telefoonverkeersrekening, De Ingenieur 52 (1937), pp. E15-E25.

[41] LASDON, L.S., Optimization Theory for Large Systems, Macmillan, New York, 1970.

[42] LENT, A., Maximum entropy and multiplicative ART, pp. 249-257 in, Shaw, R. (Editor), Image Analysis and Evaluation, SPSE Conference Proceedings, Toronto, Canada (1976).

[43] LENT, A. and CENSOR, Y., Extensions of Hildreth's row-generation method for quadratic programming, Technical Report MIPG8, Medical Image Processing Group, Dept. of Computer Science, SUNY at Buffalo, Amherst, N.Y., Sept. 1978.

[44] LUENBERGER, D.G., Optimization by Vector Space Methods Wiley, New York, 1969.

[45] MAGNANTI, T.L., Optimization for Sparse Systems, pp. 147-176 in: Bunch, J.R. and Rose, D.J. (Editors), Sparse Matrix Computations, Academic Press, 1976.

[46] McCORMICK, S.F., The methods of Kaczmarz and row orthogonalization for solving linear equations and least squares problems in Hilbert space, Indiana Univ. Math. J. 26 (1977), pp. 1137-1150.

[47] MINERBO, G.N. and SANDERSON, J.G., Reconstruction of a source from a few (2 or 3) projections, Technical Report LA-6747-MS, Los Alamos Scientific Laboratories, New Mexico, March 1977.

[48] MINSKY, M. and PAPERT, S., Perceptrons, An Introduction to Computational Geometry, The MIT Press, Second rev. printing, 1972.

[49] MOTZKIN, T.S., SCHOENBERG, I.J., The relaxation method for linear inequalities, Canad. J. Math. 6 (1954), pp. 393-404.

[50] NILSSON, N.J., Learning Machines, McGraw-Hill, New York, 1965.

[51] OETTLI, W., Einzelschrittverfahren zur Losung Konvexer und Dual-Konvexer Minimierungsprobleme, Zeit. Angew. Mathematik. Mechanic. 54 (1974), pp. 343-351.

[52] OETTLI, W., Symmetric duality and a convergent sub-gradient method for discrete, linear, constrained approximation problems with arbitrary norms appearing in the objective function and in the constraints, J. Approx. Theory 14 (1975), pp. 43-50.

[53] PITTEL, B.G., Random allocation with constraints and the principle of maximum weighted entropy, Soviet Math. Dokl. 13 (1972), pp. 1713-1715.

[54] POLYAK, B.T., Minimization of unsmooth functionals, USSR Comp. Math. and Math. Phys. 9 (1969), pp. 14-29.

[55] TANABE, K., Projection method for solving a singular system of linear equations and its applications, Numer. Math. 17 (1971), pp. 203-214.

[56] TEWARSON, R.P., Projection methods for solving sparse linear systems, The Computer J. 12 (1969), pp. 78-81.

[57] WENDLER, K., Die ε-Storung Linearer und Quadratischer Optimierungsaufgaben und Ihre Anwendung auf das Hildreth-Verfahren, Unternehmensforschung 15 (1971), pp. 1-14.

[58] WU, L. S-Y., A dynamic theory for the class of games with nonempty cores, SIAM J. Appl. Math. 32 (1977), pp. 328-338.

# Lanczos and the Computation
# in Specified Intervals of the Spectrum
# of Large, Sparse Real Symmetric Matrices

## Jane Cullum* and Ralph A. Willoughby*

Abstract. We present a Lanczos tridiagonalization procedure without reorthog-onalization for computing many or even all of the eigenvalues of large, sparse, real symmetric matrices. This procedure can also be used to compute the eigenvalues in user-specified intervals . It requires the computation of eigenvalues of 2 tridiagonal matrices of order larger than the given matrix. It has worked well on many examples. Included are numerical results and arguments to justify the proposed procedure.

1.  Introduction. Lanczos tridiagonalization procedures have proved to be very effective for computing a few extreme eigenvalues (and corresponding eigenvectors) of large, sparse, real symmetric matrices. See for example, Paige [1], Cline et al [2], Cullum and Donath [3], [4], Lewis [5], Newman and Pipano [6]. There is in the literature, however, Paige [7], van Kats and van der Vorst [8], [9], empirical evidence that Lanczos procedures may in fact be capable of computing many or even all of the eigenvalues of such matrices. (We make no claims about the computation of the associated eigenvectors.) It is this phenomenon which we investigate. We offer an explanation for it, and use the relationships developed to obtain a Lanczos procedure without reorthogonalization for computing many or even all of the eigenvalues of large, sparse, real symmetric matrices. This procedure can also be used to compute all the eigenvalues within user-specified intervals.

In section 2 we outline the Lanczos tridiagonalization procedure, review some of its properties, and summarize some of the recent research on Lanczos procedures for computing eigenelements. We include an application from the physics of disordered materials. In section 3 we outline our Lanczos algorithm without reorthogonaliza-tion. It differs substantially from the Lanczos algorithms proposed in Edwards et al

---

*IBM-T. J. Watson Research Center, Yorktown Heights, NY  10598.

[10] and van Kats and van der Vorst [9]. Like the algorithms in [9] and [10], it requires the computation of eigenvalues of 2 tridiagonal matrices of order larger than the original matrix.

In section 4 we present results of numerical tests which demonstrate clearly that our Lanczos procedure can be very effective. The storage requirements are minimal; the amount of computation required depends upon the eigenvalue distribution in the given matrix.

In sections 5 and 6, adopting ideas from optimization theory and abandoning classical requirements of global orthogonality of the Lanczos vectors, we present an explanation of the observed convergence of Lanczos tridiagonalization algorithms without reorthogonalization and give plausibility arguments for the tests and estimates used in our particular implementation of such an algorithm. These arguments use the relationships between the Lanczos tridiagonalization and the conjugate gradient optimization algorithm derived in Cullum and Willoughby [11], [12].

2. Lanczos Tridiagonalization Procedures. Lanczos tridiagonalization procedures use the Lanczos recursion

(1)  $\beta_{i+1} v_{i+1} = Av_i - \alpha_i v_i - \beta_i v_{i-1}, \quad 1 \leq i \leq m$, where

(2)  $\beta_1 = 0, \ \alpha_i = v_i^T (Av_i - \beta_i v_{i-1})$ and $|\beta_{i+1}| = \| Av_i - \alpha_i v_i - \beta_i v_{i-1} \|$.

In (1) A is the given matrix of order n, and each $v_i$ is an n component vector. There are several theoretically equivalent versions of (2), see Paige [7] for a discussion of the merits of each. We have used the version that he recommends. Observe that A is not explicitly modified, only the products Ax are required. Furthermore, the recursion uses only the 2 most recently generated v-vectors.

This recursion has been used in many different ways. In Paige [1], [7], [13], Lewis [5] it was used to compute several extreme eigenvalues of A. In Cullum and Donath [3], Underwood [14], Ruhe [15], block versions of this recursion were developed and used to compute a few extreme eigenvalues and corresponding eigenvectors. In Kaplan and Gray [16], it was used to compute approximations to certain entries in the inverse of the matrix $A-\mu I$.

Lanczos procedures replace $A$ by symmetric tridiagonal matrices (in the block versions these are block tridiagonal matrices),

$$(3) \quad T_j = \begin{bmatrix} \alpha_1 & \beta_2 & & & & \\ \beta_2 & \alpha_2 & \beta_3 & & & \\ & & \ddots & \ddots & & \\ & & & & \beta_j & \\ & & & \beta_j & \alpha_j \end{bmatrix} \quad 1 \le j \le m.$$

The eigenelement procedures approximate eigenvalues of $A$ by some subset of the eigenvalues of $T_j$ for some $j$. In exact arithmetic the vectors $v_i$ are orthonormal and $T_j = V_j^T A V_j$, $1 \le j \le n$ is the projection of $A$ onto the space spanned by $V_j = [v_1,...,v_j]$. Therefore, the eigenvalues of $T_j$ are the eigenvalues of $A$ restricted to the span of $V_j$. In particular, if the starting vector $v_1$ has a non-zero projection on the eigenspace of $A$ corresponding to each distinct eigenvalue, then for some $m \le n$ the eigenvalues of $T_m$ are the distinct eigenvalues of $A$.

In practice, however, the $v_i$ are not globally orthogonal. Paige [1] demonstrated that losses in global orthogonality were due to convergence of eigenvalues of $T_j$ to eigenvalues of $A$. Paige [13] also demonstrated that local orthogonality is maintained as long as the $\beta_i$ are not too small. In fact, Paige [13], Lewis [5] observed empirically that global orthogonality was not necessary for the successful use of the Lanczos tridiagonalization to compute extreme eigenvalues of $A$. Local orthogonality, however, is necessary.

The mathematics literature has concentrated on understanding Lanczos procedures when the order $m$ of $T_m$ is smaller than that of $A$ and when near-global orthogonality of the V-vectors is maintained (with the exception of Paige [1], [7], [13]). Estimates by Kahan [17], Kahan and Parlett [18], [19] provide a means of monitoring losses of global orthogonality and for estimating the closeness of all of the eigenvalues of $T_m$ to eigenvalues of A. These uniform estimates depend inversely upon the smallest singular value of the matrix $V_m$. The estimate in Kahan and Parlett [18] is a key element of the Parlett and Scott [20] algorithm for computing several extreme eigenvalues of A and corresponding eigenvectors. Parlett and Scott [20] maintain near-global orthogonality of the V-vectors by occasionally reorthogonalizing selected Lanczos vectors $v_i$ w.r.t. converged eigenvectors of A, see [20] for details. This is similar to the limited but nonselective reorthogonalization used in Cullum and Donath [3]. In [1], [6], [3], [14] short chains of vectors were generated and only a few eigenelements were approximated. Block procedures are not suitable for computing many eigenelements because of the amount of storage and computation required. The reorthogonalization in the Parlett and Scott algorithm [20] requires the computation of some of the eigenvectors of $T_m$. Therefore, it is also not suitable for computing many eigenvalues.

If we insist that every eigenvalue of $T_m$ must approximate an eigenvalue of A, then near-global orthogonality is essential. Near-global orthogonality cannot be maintained, see Paige [1], without reorthogonalization with respect to converged eigenvectors. If we want to use the Lanczos tridiagonalization in its original form without reorthogonalization, then we must drop the requirement that all of the eigenvalues of $T_m$ approximate eigenvalues of A. Paige [7], and van Kats and van der Vorst [9], have demonstrated empirically on small matrices A that excellent approximations to every distinct eigenvalue of A can be obtained as long as local near-orthogonality is maintained and m is large enough. With $m>n$, however, spurious eigenvalues can appear, their number and locations vary with m, and the multiplicities of the eigenvalues of $T_m$ have no relationship to the multiplicities of the eigenvalues of A. In this situation, the estimates in Kahan and Parlett [17], [18] are not applicable, and a different explanation for the observed convergence must be developed. In section 5 we give an explanation for this convergence. The arguments use the relationships between the Lanczos procedure without reorthogonalization and the conjugate gradient optimization procedure, for solving $Ax=b$.

Physicists such as Edwards et al [10] have tried to use the Lanczos algorithm without reorthogonalization with m>n to compute all the eigenvalues of large, sparse real symmetric matrices that describe energy relationships in disordered lattices of atoms. The term disorder refers to the random nature associated with one of several types of atoms occupying a given site in a lattice or rectangular array of atoms. The lattice can be 1, 2, or 3-dimensional and the x, y and z vibrations can be coupled or uncoupled. The dynamic system associated with the lattice can be reduced to a linear algebraic eigenproblem, $Au=\lambda u$. Different boundary conditions in the lattice yield different matrices. Of interest is the distribution of the eigenvalues of A over its spectrum. Computation of this distribution for different boundary conditions provides insight into the physical quantities of the material being studied such as its electrical conductivity, heat capacity and thermal conductivity, Kirkpatrick and Eggarter [21].

Before proceeding we introduce some additional terminology. We define

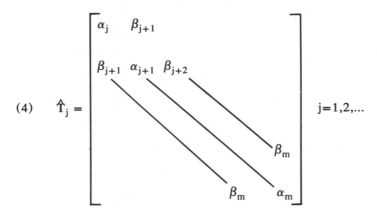

$$(4) \quad \hat{T}_j = \begin{bmatrix} \alpha_j & \beta_{j+1} & & & \\ \beta_{j+1} & \alpha_{j+1} & \beta_{j+2} & & \\ & & \ddots & & \\ & & & & \beta_m \\ & & & \beta_m & \alpha_m \end{bmatrix} \qquad j=1,2,\dots$$

We also define $a_j(\mu)$ and $\hat{a}_j(\mu)$ to be respectively the determinants of the matrices $(\mu I - T_j)$ and $(\mu I - \hat{T}_j)$. The determinants of $T_j$ and $\hat{T}_j$ are denoted by $a_j$ and $\hat{a}_j$. The eigenvalues of $T_j$ are denoted by $\mu_1^j \leq \dots \leq \mu_j^j$ and those of $\hat{T}_j$ by $\hat{\mu}_1^j \leq \dots \leq \hat{\mu}_{m-j+1}^j$. The j superscript will be omitted whenever it is clear from the context. The derivative of $a_j(\mu)$ is denoted by $a_j'(\mu)$. The distinct eigenvalues of A are denoted by $\lambda_1 \leq \dots \leq \lambda_q$ where $q \leq n$ and corresponding orthonormal eigenvectors are denoted by $z_k$, $1 \leq k \leq q$.

In the next section we describe our Lanczos procedure without reorthogonalization of the V-vectors for computing many or even all of the eigenvalues of large, real symmetric matrices. The key to the success of this algorithm is the test for determining which eigenvalues of $T_m$ are spurious.

3. <u>A Lanczos Algorithm without Reorthogonalization</u>. We describe a Lanczos algorithm with no reorthogonalization for computing many or all of the eigenvalues of a large, real symmetric matrix A. This procedure uses recursion (1); it does not explicitly alter A. It requires the fast, accurate computation of matrix-vector products Av. Sparsity is not required, but if the matrix is sparse in the sense that only $O(n)$ operations are required to compute Av, then the required speed and accuracy are typically available. As programmed, storage for 5 and 1/2 double precision vectors of length mmax is required where mmax is the maximal order of $T_m$ being considered. In section 4 we give an example that demonstrates how this procedure can be used to compute eigenvalues in user-specified intervals.

Eigenvalues of A are approximated by a subset of the eigenvalues of $T_m$ for some m. The computational efficiency of this procedure depends upon the efficiency of the algorithm used to compute the eigenvalues of the tridiagonal matrices used, and the distribution of the eigenvalues in A. As implemented, the programs use the EISPAK [22] subroutines IMTQL1 and BISECT. Eigenvectors are not computed. Our Lanczos procedure compares eigenvalues from 2 tridiagonal matrices. Therefore, it is essential that the eigenvalues of the tridiagonal matrices be computed reasonably accurately. If they are not computed accurately nonspurious eigenvalues may be labelled spurious erroneously.

First we describe the procedure and then explain how it differs from the procedures proposed in Edwards et al [10] and van Kats and van der Vorst [9]. We note that this procedure cannot determine the multiplicities of the eigenvalues of A. We are currently testing a related procedure that could be used to determine which eigenvalues are multiple.

Please refer to the flow chart in Figure 1. The unit starting vector $v_1$ is generated randomly. A basic assumption is that the starting vector has a non-trivial

FIGURE 1. FLOWCHART - LANCZOS PROCEDURE

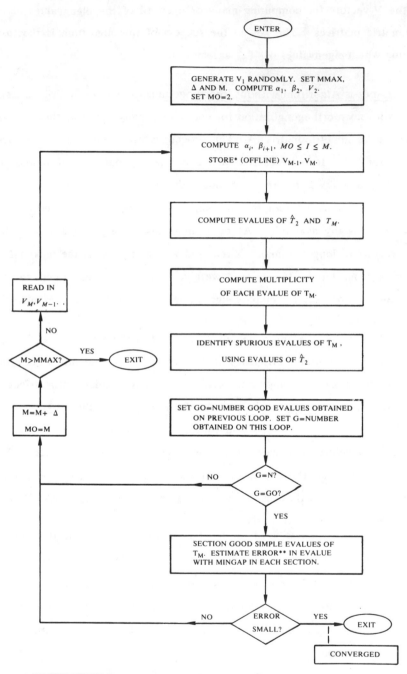

*ALTERNATIVELY, $T_{MMAX}$ COULD BE GENERATED INITIALLY AND STORED.

**THE TEST ON G COULD BE INTERCHANGED WITH THE ERROR ESTIMATION.

projection on the invariant subspace of each of the desired eigenvalues. Using $v_1$ and recursion (1), sequences of scalars $\alpha_i$, $\beta_{i+1}$, i=1,2,...,m are generated. The user must specify m, the order of $T_m$. If the user has no information about the eigenvalue distribution in the matrix A, then setting m=2n or 3n is recommended. As indicated by the examples tested, if the eigenvalues of the matrix A have gaps not too much smaller than the average gap = spread/n (isolated pairs of close eigenvalues are allowed), then often all the eigenvalues can be computed successfully by m=2n. See section 4 for details on convergence for various distributions of eigenvalues. The vectors $v_i$ are not stored with the exception of $v_{m-1}$ and $v_m$ which may be stored off-line. During the computations a history is created. The $\alpha_i$, $\beta_i$, the vectors $v_m$, $v_{m-1}$ and the eigenvalues of $\hat{T}_2$ and $T_m$ are stored so that if the user decides to recompute using either the same or a larger m, the program can start where it left off on the most recent run.

The procedure outlined in Figure 1 uses IMTQL1 in EISPAK [22] to compute the eigenvalues of $T_m$ and of $\hat{T}_2$, the matrix of order m-1 obtained from $T_m$ by deleting the first row and column. In section 4 we give an example of the use of BISECT for this computation when either the user wants only the eigenvalues in some subinterval of the spectrum or the eigenvalues in a cluster must be computed for a larger value of m in order to achieve the desired convergence. The multiplicity of each of the eigenvalues in $T_m$ is determined using the tolerance,

(5)     $toln = m \times macheps \times max\{|\mu_k|, 1 \leq k \leq m\},$

which is an estimate of the accuracy achievable in the IMTQL1 subroutine. In (5) macheps is the relative precision of the machine arithmetic. For the IBM 370/168 we set macheps $= 10^{-15}$. Observe that theoretically $T_m$ cannot have multiple eigenvalues since all the $\beta_k$ are non-zero. However, numerically it can and does have multiple eigenvalues. Each multiple eigenvalue of $T_m$ is accepted as an approximate eigenvalue of A. (See Lemma 1 for a justification of this.)

Spurious eigenvalues are identified by comparing the simple eigenvalues of $T_m$ with the eigenvalues in $\hat{T}_2$. If for any simple eigenvalue $\mu_k$ of $T_m$ there exists an eigenvalue $\hat{\mu}_j$ of $\hat{T}_2$ such that

(6)     $|\hat{\mu}_j - \mu_k| < \text{toln}$ ,

then $\mu_k$ is labelled spurious and not an approximate eigenvalue of A. Spurious eigenvalues are indexed by 0, nonspurious eigenvalues are indexed by their multiplicities. An example of the use of this identification test as well as a spurious/multiplicity pattern for a test matrix of order n=465 with m=800 is given in Cullum and Willoughby [23].

The number of good or nonspurious eigenvalues g obtained on the current loop is compared with the order of A and with the number go obtained on the previous loop, (if any). Estimates of the errors in the good eigenvalues are also made using the bound in Lemma 4. If g=n or g=go and the error estimates are smaller than a user specified tolerance, the procedure terminates. This can be modified to use only one or the other of these tests for convergence. In particular if fewer than all of the distinct eigenvalues of A are desired, then just the error estimates could be used and only on the desired range of eigenvalues.

The error estimates utilize a subroutine that works separately with the mantissa and the exponent of each number. This is necessary because the products involved can easily become large (small) enough to cause overflow (underflow). We can reduce the amount of work required to compute these estimates by using the empirical observation that an eigenvalue $\lambda_k$ converges according to its relative location in the spectrum of A, extreme or interior, the associated minimal gap,

(7)     $\min [\lambda_{k+1} - \lambda_k, \lambda_k - \lambda_{k-1}]$,

and whether or not it is in a cluster. This agrees with the theoretical relationships discussed in [11], [12]. The accuracy of the eigenvalue approximations within a given portion of the spectrum can be estimated by the error in the eigenvalue with the minimal gap on that section.

Our algorithm differs from those proposed in Edwards et al [10] and van Kats and van der Vorst [9] in several ways. First, we use $\hat{T}_2$ and $T_m$; they use $T_{m-1}$ and $T_m$. Second, they accept an eigenvalue as good if it is an eigenvalue of both

$T_{m-1}$ and of $T_m$. We <u>reject</u> any simple eigenvalue of $T_m$ that is also an eigenvalue of $\hat{T}_2$. The matrices $T_{m-1}$ and $T_m$ are very similar, and the question arises as to what we mean when we say they have the same eigenvalues. Depending on the tolerance set in such a test, one can accept spurious eigenvalues as good. If one sets this tolerance to toln in (5), then this ambiguity disappears. However, then one obtains an approximation to an eigenvalue of A only after that approximation is accurate to toln. $\hat{T}_2$ and $T_m$ are not similar. The arguments in section 6 indicate that any simple eigenvalue of $T_m$ that is pathologically close to an eigenvalue of $\hat{T}_2$ is spurious. This test requires the accurate computation of the eigenvalues of $\hat{T}_2$ and $T_m$ and use of the tolerance in (5). Since we are rejecting spurious eigenvalues rather than trying to identify good ones, we obtain approximations to eigenvalues of A when these approximations are just starting to converge. This allows us to have a convergence test that is tied to the number of good eigenvalues obtained to date.

The ability of this procedure to identify spurious eigenvalues is limited by the current computational accuracy in IMTQL1 to matrices with gaps greater than single precision, $10^{-6}$. Moreover, if the spread is large and there are large differences in the magnitudes of the eigenvalues of A, then the tolerance in (5) will be large and may cause some confusion in identifying the eigenvalues of small magnitude. Note that toln in (5) depends only on the eigenvalue of largest magnitude and as programmed toln is used as an absolute tolerance. For this reason it is suggested that the user save the eigenvalues of $\hat{T}_2$ and $T_m$ in a file, at least those portions of the eigenvalues that are largest and smallest in magnitude. If the spread is large and there are large differences in the magnitudes of the eigenvalues, then some of the eigenvalues that are smallest in magnitude may have been classified erroneously as spurious. In this situation the user should rerun the spurious test on the small magnitude portion of the spectrum using the tolerance

$$(8) \quad \text{toln} = m \times \text{macheps} .$$

Unless we obtain as many nonspurious eigenvalues as the order of A, we cannot be absolutely certain that we have obtained all the distinct eigenvalues of A. There is always the possibility that our starting vector $v_1$ was deficient in one or more of the eigenvectors of A. There is no way to test for this deficiency other than changing the starting vector and rerunning the procedure.

This procedure, as currently programmed, uses 5 and 1/2 double precision vectors of length mmax plus whatever additional storage the user requires to generate the matrix-vector products Ax. For large matrices, the storage requirements are very small compared with the standard EISPAK algorithms which require at least $n(n+1)/2$ double precision words. For example, if $n=1000$ and mmax=6000, then the proposed algorithm requires approximately $2.64 \times 10^5$ bytes, while the standard algorithm would require approximately $4 \times 10^6$ bytes, 15 times as much storage. More typically, mmax $\leq$ 3000 and the preceding factor is greater than 30. Thus, this procedure allows us to work with much larger matrices than the standard procedures do and to take advantage of any sparsity or patterns in the given matrix.

The significant parts of the computation are summarized below in Table 1.

TABLE 1. Computational Cost

|  | No. Multiplications |
| --- | --- |
| Generation of $T_m$ | $mn(q+2)$ |
| Compute Eigenvalues of $\hat{T}_2$ and $T_m$ | $2m^2r$ |
| Error Estimates | $2mn$ (n subroutine calls) |

In Table 1 the following are assumed. Each matrix-vector product Ax requires qn multiplications. On the average r internal iterations in IMTQL1 are required to compute one eigenvalue of $T_m$, and the count assumes that <u>all</u> the eigenvalues of $\hat{T}_2$ and $T_m$ are computed. The eigenvalue computations are currently taking the largest fraction of the total computational time. The error estimates are expensive because they require repeated calls to a subroutine which transforms the quantities involved to their mantissa, exponent form. The cost in Table 1 is for computing these estimates for every eigenvalue.

Savings in computational costs can be realized by using the EISPAK [22] routine, BISECT , in addition to or instead of IMTQL1. IMTQL1 could be used initially on a reasonably small value of  m, for example  2n or  3n, to compute all of the eigenvalues of  $\hat{T}_2$  and all of the eigenvalues of  $T_m$. Estimates of convergence on the good eigenvalues could then be computed. Indications of clusters of eigenvalues, if any, will be apparent from the size of these errors which can be used to identify those parts of the spectrum of  A which have not been computed accurately. The  $T_m$  generation could then be continued to a large value of  m, say  m=6n, and then BISECT could be used to compute only those portions of the spectrums of  $\hat{T}_2$  and  $T_m$  that had not converged sufficiently for the first choice of m.  BISECT requires less computational effort than IMTQL1 if only a small fraction of the eigenvalues of  A  remain to be computed.

Summarizing, we have replaced the problem of computing eigenvalues of a large, sparse but unstructured matrix by the problem of computing eigenvalues of 2 larger, highly structured, in fact, tridiagonal matrices.  We note that if we require only the eigenvalues in some portion of the spectrum of  A, then we need compute only the eigenvalues in the same portions of the spectrums of  $T_m$  and  $\hat{T}_2$.  If, however, we want to use our estimates of the accuracy of the computed eigenvalues, then we would have to compute all of the eigenvalues of at least  $\hat{T}_2$.  Alternatively, we could enlarge  $T_m$  and then compute the eigenvalues of this larger matrix in the relevant portions of the spectrum and determine by comparison if in fact they have converged.  Thus, the procedure could be used to compute the interior eigenvalues of the given matrix within a specified interval.

4.  Numerical Results.  Tests of the algorithm described in section 3 have been made on several families of matrices:  Poisson-type matrices with and without normal derivative boundary conditions, matrices related to disordered systems, Kirkpatrick and Eggarter [21], diagonal matrices and some smaller order matrices that have appeared in the literature, van Kats and van der Vorst [8], [9].  Brief descriptions of the Poisson and disordered test matrices are given below.  Summariz-

ing the results of such tests is difficult. In each case considered with the minimal gaps greater than $10^{-7}$, the algorithm has successfully computed all of the distinct eigenvalues of the original matrix. In a very stiff diagonal problem with $\lambda_1=0$, $\lambda_2=10^{-8}$, and $\lambda_n=10^5$ the identification test as given in (6) failed to recognize the smaller eigenvalues. As stated earlier, in such a situation where there is a large difference in magnitude in the eigenvalues, the absolute tolerance in (8) should be used on the smaller eigenvalues. The problem in that case was not that the procedure had not computed all of the eigenvalues of A, but that the tolerance in the test for spuriousness was too large so several of the eigenvalues of smallest magnitude were classified incorrectly as spurious. It is not possible to describe all of the examples in detail. We plan to present this elsewhere. We present 3 specific examples to illustrate the behavior of the algorithm, a Poisson-matrix of order 992 with zero, normal derivative boundary conditions, POIS992; a disordered matrix of order 992, KIRK992; and a Poisson-matrix of order 1008 with zero boundary conditions, POIS1008. The third example, POIS1008, is also discussed in Cullum and Willoughby [23]. In the first 2 examples POIS992 and KIRK992 the eigenvalues and eigenvectors were not known a priori. In the third example the eigenelements were known.

Poisson-type matrices. We consider the Laplace equation on a rectangle,

$$(9) \qquad u_{xx} + u_{yy} = \lambda'u, \quad R = \{(x,y) \mid 0 \leq x \leq X, \ 0 \leq y \leq Y\}.$$

We consider two types of boundary conditions: (Z) $u=0$ on the boundary of R, and (N) $\partial u/\partial n = 0$ on the boundary except $u=0$ when $y=0$. We replace (9) by a scaled discretization, $Ax = \lambda x$. For given $\Delta x$ and $\Delta y$ we choose $k_x$ and $\ell_y$ so that $X = k_x \Delta x$ and $Y = \ell_y \Delta y$. The order of A is $N = KL$ where for boundary conditions (Z) $K = k_x - 1$ and $L = \ell_y - 1$. For conditions (N) $K = k_x + 1$ and $L = \ell_y$. A is block tridiagonal, each block is $K \times K$ and there are L blocks B down the diagonal.

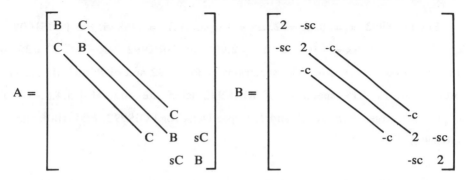

and $C = -(1-c)I$ where $c = 1/(1 + (\Delta x/\Delta y)^2)$. For (Z) $s=1$ and for (N) $s = \sqrt{2}$. Note that we have scaled the ordinary discretization of the Laplace operator by $c(\Delta x)^2$. For boundary conditions (Z) the eigenvalues of A, with $1 \leq I \leq K$, $1 \leq J \leq L$, are $\lambda(I,J) = 2[1 - c \cos(\pi I/k_x) - (1-c) \cos(\pi J/\ell_y)]$. For boundary conditions (N) the eigenvalues are not known a priori. By varying the value of c we have obtained matrices with many different eigenvalue distributions.

Diagonally-disordered matrices. The matrices we considered arise in the study of two-dimensional arrays of atoms, NX by NY, in disordered systems Kirkpatrick and Eggarter [21]. We considered the diagonally-disordered case. The matrix A has order $N = NX \times NY$. If NX and NY are relatively prime, then all of the eigenvalues of A are distinct. A is almost block tridiagonal. Each block is NX $\times$ NX, and there are NY blocks B down the diagonal.

$$A = \begin{bmatrix} B & C & & & & C \\ C & B & & & & \\ & & \ddots & & & \\ & & & & & C \\ & & & & & \\ C & & & & C & B \end{bmatrix} \qquad B = \begin{bmatrix} x & 1 & & & & 1 \\ 1 & x & 1 & & & \\ & & \ddots & & & \\ & & & & & 1 \\ & & & & & \\ 1 & & & & 1 & x \end{bmatrix}$$

and C is the unit matrix of order NX. We have denoted the diagonal entries by x's, these are randomly generated numbers. A scaling parameter bounds the magnitudes of these disorder terms. All of the nonzero, off-diagonal elements have value 1 and their pattern is that of the discretization of Laplace's equation using nearest (x,y) neighbors and with doubly periodic boundary conditions.

For POIS992 $k_x=30$, $\ell_y=32$, $\Delta x = 1.0$ and $\Delta y = .96875$. For POIS1008 $k_x = 57$, $\ell_y = 19$, $\Delta x = 1.0$ and $\Delta y = 2.99$. For KIRK992 NX= 31, NY=32, and the scaling parameter is 6. The spectrums of POIS992 and of POIS1008 are in the interval (0,4). The spectrum of KIRK992 is in the interval (-5.4,5.4). The $|\beta_i|$ were larger than .58, .5 and 1. respectively, for POIS992, POIS1008 and KIRK992.

Table 2 summarizes the observed convergence of the eigenvalues of these three examples as m is increased. For each value of m considered we have listed the number of eigenvalues of A being approximated, the number of these eigenvalues which are accurate to at least 10 significant digits and the number which are accurate to at least 6 significant digits. For examples 1 and 2 these numbers are actually lower bounds. For these two examples we used the error estimates E in Equation (30) and not the true error to obtain the numbers used in Table 2. These estimates are conservative and in the cases where comparisons of the true error with the estimated error have been made the eigenvalues have always had several digits more accuracy than indicated by the estimates when the estimates were still reasonably large $\geq 10^{-6}$. For Example 3 we computed the true relative error.

Example 3, POIS1008, exhibited the most rapid convergence. By $m=1.75n=1764$ all of the eigenvalues of A have been computed to at least 10 significant digits. The diagonally-disordered matrices also exhibited rapid convergence, by $m=2n$, the estimates indicate that all of the eigenvalues are good to 8 digits. They are in fact good to at least 10 significant digits. Similar convergence has been observed on other diagonally-disordered matrices of order 567, 927 and 1505, respectively.

Example 1, POIS992, exhibits slower convergence. Observe that by $m=2.5n$, 974 eigenvalues are being approximated, 919 of these are good to at least 6 digits. These numbers were obtained from the estimates, not by computing the true error. Several eigenvalues are not being approximated and it is not until somewhere between 4n and 5n that every eigenvalue is represented. Observe, however, that only 18 eigenvalues are missing at $m=2.5n$.

TABLE 2. Convergence Patterns.

| Matrix | Order of $T_m$ | Eigenvalues Approximated | Accuracy $\geq 10$ Digits | Accuracy $\geq 6$ Digits |
|--------|--------|--------|--------|--------|
| POIS992 | | | | |
| | 1984(2n) | 945 | 667 | 774 |
| | 2480(2.5n) | 974 | 830 | 919 |
| | 2976(3n) | 981 | 896 | 956 |
| | 3472(3.5n) | 986 | 917 | 958 |
| | 4000(4n) | 988 | 943 | 958 |
| | 4960(5n) | 992 | 949 | 992 |
| | 5952(6n) | 992 | 992 | |
| KIRK992 | | | | |
| | 496(.5n) | 432 | 67 | 88 |
| | 922(n) | 728 | 215 | 256 |
| | 1488(1.5n) | 921 | 467 | 500 |
| | 1984(2n) | 992 | 957 | 992 ($\geq 8$ Digits) |
| | 2480(2.5n) | 992 | 992 | |
| POIS1008 | | | | |
| | 504(.5n) | 478 | 62 | 70($\geq 8$ Digits) |
| | 756(.75n) | 681 | 113 | 123($\geq 8$ Digits) |
| | 1008(n) | 867 | 171 | 307 |
| | 1260(1.25n) | 989 | 852 | 878($\geq 8$ Digits) |
| | 1512(1.5n) | 1006 | 991 | 997($\geq 8$ Digits) |
| | 1764(1.75n) | 1008 | 1008 | |

The difference in convergence between POIS992 and KIRK992 can be explained by considering Tables 3 and 4. Table 3 gives the distribution of gap intervals for each of these matrices. We compute the minimal gap, see Equation (7), for each eigenvalue of A and then compute the number of eigenvalues whose gap intervals lie within a given range of the average gap = spread/n. In Table 3 gap intervals numbered $\geq 7$ correspond to gaps $\geq$ average gap. Gap intervals numbered $\geq 5$ are $\geq$ average gap/10, numbered $\geq 3$ are $\geq$ average gap/100, etc. Moreover, we have arbitrarily divided the spectrums into interior and extreme eigenvalues by a simple cardinality rule. We order the spectrums by increasing size. In each spectrum eigenvalues numbered 1 to $3n/11$ and $(1-3/11)n$ to n are labelled extreme; those numbered $3n/11$ to $(1-3/11)n$ are labelled interior. For the purposes of comparison this makes sense only if the two matrices have approximately the same number of eigenvalues in the corresponding intervals of the spectrums. This is approximately true for these two examples.

TABLE 3. Eigenvalues in Gap Intervals, POIS992 and KIRK992.

| Gap Interval | 1 | 2 | 3 | 4 | 5 | 6 | 7 | 8 |
|---|---|---|---|---|---|---|---|---|
| POIS992 (Average Gap = $4 \times 10^{-3}$) | | | | | | | | |
| % of Evalues | 0. | 1.4 | 4.4 | 13.5 | 17.7 | 49.2 | 13.2 | 0.4 |
| No. Extreme | 0 | 8 | 20 | 60 | 84 | 266 | 99 | 4 |
| No. Interior | 0 | 6 | 24 | 74 | 92 | 222 | 33 | 0 |
| KIRK992 (Average Gap = $1.1 \times 10^{-2}$) | | | | | | | | |
| % of Evalues | 0. | 0.2 | 0.4 | 1.8 | 53.6 | 31.6 | 13.1 | 0.2 |
| No. Extreme | 0 | 2 | 4 | 12 | 253 | 167 | 102 | 2 |
| No. Interior | 0 | 0 | 0 | 6 | 269 | 147 | 28 | 0 |

In Table 3 we see that only 6 interior eigenvalues of KIRK992 have gaps less than 1/10 of the average gap. In POIS992 104 interior eigenvalues have gaps less than this. This indicates a potential difference in the convergence of the two matrices. However, it is quite possible for a matrix to have eigenvalues with small gaps and to still get good convergence. Isolated pairs of eigenvalues can be very close together and this will not affect the rapidity of convergence.

We have to go to Table 4 to understand the convergence observed. In Table 4 we have listed the gap intervals for the middle of the spectrum for both examples. In the middle of the spectrum of Kirk992 we see there are no eigenvalues with gaps $\leq$ average gap/10. For POIS992 however, there are long chains of eigenvalues with small gaps on either side of the middle of the spectrum. The difficulties with the convergence occur in these two chains. Every other eigenvalue of POIS992 is present at $m=2.5n$. The last ones to converge are in the clusters of 2's in Table 4.

TABLE 4. Gap Intervals, Eigenvalues 421 to 560, POIS992 and KIRK992.

Matrix Eval. Gap Intervals

POIS992

| 421 | 6 | 4 | 4 | 4 | 6 | 6 | 6 | 3 | 3 | 5 | 6 | 6 | 6 | 6 | 6 | 6 | 6 | 6 | 4 | 4 |
|-----|---|---|---|---|---|---|---|---|---|---|---|---|---|---|---|---|---|---|---|---|
| 441 | 5 | 5 | 6 | 6 | 6 | 6 | 6 | 6 | 6 | 3 | 3 | 6 | 6 | 5 | 5 | 7 | 3 | 3 | 6 | 6 |
| 461 | 6 | 6 | 2 | 2 | 2 | 4 | 4 | 5 | 4 | 4 | 4 | 5 | 4 | 3 | 3 | 4 | 4 | 3 | 3 | 6 |
| 481 | 5 | 5 | 4 | 4 | 5 | 5 | 6 | 6 | 6 | 7 | 6 | 6 | 6 | 6 | 6 | 5 | 5 | 6 | 6 | 6 |
| 501 | 6 | 6 | 7 | 6 | 6 | 6 | 5 | 5 | 4 | 4 | 5 | 5 | 6 | 3 | 3 | 4 | 4 | 3 | 3 | 4 |
| 521 | 5 | 4 | 4 | 4 | 5 | 4 | 4 | 2 | 2 | 2 | 6 | 6 | 6 | 6 | 3 | 3 | 7 | 5 | 5 | 6 |
| 541 | 6 | 3 | 3 | 6 | 6 | 6 | 6 | 6 | 6 | 6 | 5 | 5 | 4 | 4 | 6 | 6 | 6 | 6 | 6 | 6 |

KIRK992

| 421 | 5 | 5 | 5 | 5 | 7 | 6 | 5 | 5 | 5 | 5 | 5 | 5 | 7 | 5 | 5 | 5 | 5 | 5 | 5 | 5 |
|-----|---|---|---|---|---|---|---|---|---|---|---|---|---|---|---|---|---|---|---|---|
| 441 | 5 | 5 | 5 | 5 | 5 | 5 | 5 | 5 | 5 | 6 | 6 | 6 | 6 | 5 | 5 | 5 | 6 | 6 | 5 | 5 |
| 461 | 6 | 6 | 6 | 7 | 6 | 5 | 5 | 5 | 5 | 6 | 6 | 6 | 5 | 5 | 6 | 6 | 6 | 5 | 5 | 5 |
| 481 | 6 | 6 | 5 | 5 | 6 | 5 | 5 | 5 | 5 | 5 | 5 | 5 | 5 | 5 | 7 | 7 | 5 | 5 | 6 | 6 |
| 501 | 5 | 5 | 5 | 5 | 5 | 5 | 5 | 7 | 7 | 6 | 6 | 5 | 5 | 6 | 7 | 7 | 5 | 5 | 5 | 6 |
| 521 | 5 | 5 | 6 | 5 | 5 | 5 | 6 | 5 | 5 | 5 | 5 | 6 | 5 | 5 | 5 | 5 | 5 | 5 | 5 | 5 |
| 541 | 5 | 7 | 5 | 5 | 6 | 6 | 6 | 6 | 6 | 5 | 5 | 5 | 6 | 5 | 5 | 5 | 5 | 7 | 7 | 5 |

Thus, the difficulties with POIS992 are due to the clusters of eigenvalues in the middle of the spectrum. To compute these few eigenvalues accurately we had to increase m to 5n. In practice, however, when only a small fraction of the desired eigenvalues have not yet been computed accurately, we do not want to have to compute all of the eigenvalues of these large matrices. We can, however, do the following. For a given matrix use the procedure in section 3 for a relatively small value of m, m=2.5n or 3n. Compute all of the eigenvalues of $T_m$ and $\hat{T}_2$ for this m. Compute the error estimates, either E or E', in Equation (30) for the simple good eigenvalues of $T_m$. Use these estimates to identify those portions of the spectrum of A which have not converged sufficiently or are not being approximated. Numerical experience indicates that these error estimates are very effective for doing this identification. Using a large value of m, 5n or 6n, use BISECT to compute all of the eigenvalues of $T_m$ and of $\hat{T}_2$ in those intervals of the spectrum that have been identified as requiring a larger value of m to achieve convergence. In Table 5 we list one portion of a BISECT computation for POIS992 on one of the intervals in the spectrum corresponding to the underlined section of Table 4. Using these computed eigenvalues we apply the same identification test used in section 3 to identify the eigenvalues of $T_m$ that approximate eigenvalues of A. We must choose the subintervals submitted to BISECT slightly larger than indicated by our tests on the estimates, and then disregard the eigenvalues that are too close to the end points of these subintervals when we are applying our identification test. Every eigenvalue of POIS992 in this subinterval was computed and identified correctly. The amount of computation required by BISECT is a small fraction of the computation required by IMTQL1.

If BISECT is used then the error estimates in (30) cannot be used unless all of the eigenvalues of $\hat{T}_2$ are computed. Alternatively, one can increase m and recompute the eigenvalues in the relevant intervals, checking to see if the number of eigenvalues and their values have not changed with the increase in m. BISECT is particularly appropriate if the user requires only the eigenvalues in a given interval. In this situation the initial phase, using IMTQL1 to determine which intervals to supply to BISECT, can be omitted and the BISECT computation can be performed for 2 large values of m.

TABLE 5.  Compute Interior Eigenvalues of POIS992, Using BISECT.

| Classification | No. | Eigenvalue $T_m$ | No. | Eigenvalue $\hat{T}_2$ |
|---|---|---|---|---|
| Accept | 40 | 1.944148440745763 | 39 | 1.942467578639830 |
| Repeat | 41 | 1.944148440745763 | 40 | 1.944148440745763 |
| Accept | 42 | 1.946284978373996 | 41 | 1.944310777494893 |
| Repeat | 43 | 1.946284978373996 | 42 | 1.946284978373996 |
| Reject | 44 | <u>1.947690935703804</u> | 43 | 1.947077133213947 |
| Accept | 45 | 1.947690935814651 | 44 | <u>1.947690935703804</u> |
| Reject | 46 | <u>1.948757755271330</u> | 45 | 1.948041036933613 |
| Accept | 47 | 1.949375014096254 | 46 | <u>1.948757755271330</u> |
| Accept | 48 | 1.949414050689150 | 47 | 1.949396172952957 |
| Accept | 49 | 1.949445027114972 | 48 | 1.949438453783388 |
| Accept | 50 | 1.949823523784888 | 49 | 1.949526208867054 |
| Accept | 51 | 1.950074429184259 | 50 | 1.949955576612031 |
| Accept | 52 | 1.950621088818346 | 51 | 1.950370910518832 |
| Accept | 53 | 1.951482491311509 | 52 | 1.951104000592342 |
| Accept | 54 | 1.951627326668485 | 53 | 1.951484905734591 |
| Accept | 55 | 1.951871182507160 | 54 | 1.951869128221077 |
| Accept | 56 | 1.952702769093079 | 55 | 1.952536469448453 |
| Accept | 57 | 1.953588228107688 | 56 | 1.952773211744211 |
| Accept | 58 | 1.953711627151370 | 57 | 1.953631842390134 |
| Accept | 59 | 1.953754454283900 | 58 | 1.953720951969298 |

Table 6 exhibits the interesting stability of the identification test that uses $\hat{T}_2$. We consider a diagonally-disordered matrix of order 567, KIRK567. We generated $T_m$ with m=2n=1134. At this value of m all of the eigenvalues of KIRK567 have been computed to at least 10 significant digits. We then introduced random error into the $\alpha_i$ and $\beta_{i+1}$, $1 \leq i \leq m$, generating a perturbed tridiagonal matrix which we call $T_m(R)$. The random error was uniformly distributed and bounded by $10^{-5}$.

TABLE 6.  KIRK567 with Random Error of Order $10^{-5}$ in $\alpha$ and $\beta$.

Computed Eigenvalues of KIRK567

| 63 | -3.10662964842891398 | 64 | -3.06096975251351244 |
| 65 | -3.04878585184903628 | 66 | -3.03044306158149701 |
| 67 | -3.00982781816963341 | 68 | -2.99624257792801796 |
| 69 | -2.96969176086940623 | 70 | -2.95163671787189208 |
| 71 | -2.94708392952897102 | 72 | -2.93413967970963085 |

| Classification | Eigenvalue of $T_m(R)$ | Eigenvalue of $\hat{T}_2(R)$ |
| --- | --- | --- |
|  | -3.12566860307314 | -3.12554924156651 |
| reject | -3.10704920957043 | -3.10704920957042 |
| accept | -3.10663020890593 | -3.10662803517152 |
| reject | -3.10662803517112 | -3.10610908676702 |
| accept | -3.06096972982092 | -3.06096804060017 |
| reject | -3.06096804059977 | -3.05985495278045 |
| accept | -3.04878517667712 | -3.04878413992961 |
| reject | -3.04878413992931 | -3.04821655312678 |
| reject | -3.03044428965363 | -3.03044428965398 |
| accept | -3.03044304821839 | -3.02958208044722 |
| reject | -3.00982802007181 | -3.00982802007220 |
| accept | -3.00982705928303 | -3.00687412579754 |
| accept | -2.99624265127813 | -2.99624154546542 |
| reject | -2.99624154546519 | -2.99621735156353 |
| reject | -2.96969222493124 | -2.96969222493161 |
| accept | -2.96969194860352 | -2.96708587627313 |
| reject | -2.95163764274343 | -2.95163764274364 |
| accept | -2.95163667183916 | -2.95098909337554 |
| accept | -2.94708401698687 | -2.94708392832421 |
| reject | -2.94708392832392 | -2.93470624427092 |

In Table 6 we first list some of the eigenvalues of KIRK567. Then we give a small portion of the corresponding eigenvalues of $T_m(R)$ and $\hat{T}_2(R)$. Using the identification test we label each eigenvalue of $T_m(R)$ as either an approximation to an eigenvalue of A (accept) or as spurious (reject). Duplicate eigenvalues are labelled as repeats. Observe that all of the eigenvalues of A are computed to within the accuracy of the random perturbation introduced. Observe also that the spurious eigenvalues, those occurring in both $T_m(R)$ and in $\hat{T}_2(R)$, appear in both sets of eigenvalues to the accuracy of the IMTQL1 computations. The number and locations of the spurious eigenvalues changed with the random perturbations, but the fact that these eigenvalues can be identified by our test did not change. We observe that the eigenvalues of $T_m$ that were multiple split into distinct eigenvalues when the random perturbations were introduced. The eigenvalues of $\hat{T}_2$ also split, but in such a way that numerically they matched many of the eigenvalues of $T_m(R)$.

This section demonstrates that this type of procedure can be very effective especially in conjunction with a strategy that selects a relatively small m, m=2n or 3n, initially and then uses BISECT on those portions of the spectrum that have not yet converged. In the next two sections we present lemmas, theorems, and plausibility arguments to justify the Lanczos algorithm described in sections 3 and 4.

5. Why Should This Algorithm Work? We summarize several fundamental results of Paige [1], [13], and several important relationships that are valid for tridiagonal matrices. Following this we summarize the results obtained in [11] relating the Lanczos tridiagonalization and the conjugate gradient procedures. We then relate these results to the Lanczos algorithm given in section 3.

Since we will be using conjugate gradients, we must assume that the given matrix A is positive definite. For eigenvalue applications, at least theoretically, the vectors generated by (1) are the same for all matrices $A + \tau I$ for any shift $\tau$. In practice, these vectors vary with $\tau$. However, the amount of variation is controlled if the $v_i$ are locally, nearly orthogonal. Paige [13] has shown that as long as $|\beta|_{min} \equiv min\ \{|\beta_i|, 1 \leq i \leq m\}$ is not too small the Lanczos vectors are locally nearly-orthogonal.

In matrix form recursion (1) is

(10) $\quad AV_m = V_m T_m + \beta_{m+1} v_{m+1} e_m^T$

where $e_m^T = (0,0,..,0,1)$ is the mth coordinate vector. For each eigenvalue $\mu_j^m$ of $T_m$ let $u^{mj}$ be an associated eigenvector with ith component $u_i^{mj}$. We assume that the $u^{mj}$ are orthonormal. We omit the superscripts m and j whenever they are clear from the context. From Wilkinson [24] and (10) we have that given any eigenvalue $\mu$ of T and corresponding eigenvector u, there exists an eigenvalue $\lambda$ of A such that

(11) $\quad |\lambda - \mu| \leq |\beta_{m+1} u_m| / \|Vu\|.$

We call $z = Vu$ the Ritz vector corresponding to u.

**Theorem 1. Paige [1]. Given** k < m **and an eigenvalue** $\mu_s^k$ **of** $T_k$ **with corresponding eigenvector** $u^{ks}$, **then**

(12) $\quad \min \{|\mu_j^m - \mu_s^k|, 1 \leq j \leq m\} \leq |\beta_{k+1} u_k^{ks}|.$

Theorem 1 states that any eigenvalue of $T_k$ whose kth component is very small is also an eigenvalue of $T_m$ for any m > k. Thus once an eigenvalue $\mu_s^k$ of $T_k$ converges, i.e. $u_k^{ks}$ is very small, then that eigenvalue will be an eigenvalue of all succeeding $T_m$. But such eigenvalues are eigenvalues of A. Therefore, Theorem 1 tells us that if we choose an m larger than necessary to get the desired eigenvalues, there is no significant resulting loss in accuracy. It will, of course, cost us computationally. Thus, there is a great deal of freedom in our choice of m.

**Theorem 2. Paige [1]. Let the eigenvalues** $\mu_j$, k−1 ≤ j ≤ k + 1, **of** $T_m$ **satisfy for some** b > 0

(13) $\quad \mu_{k-1} + b < \mu_k < \mu_{k+1} - b,$

**and let** $u^i$ **be corresponding orthonormal eigenvectors. Then if b is not pathologically small,**

(14) $\quad \|Vu^k\| > .8.$

Thus, Theorem 2 states that $|\beta_{m+1} u_m^k|$ is a good estimate of the error in the associated eigenvalue $\mu_k^m$ when b in (13) is not pathologically small. We will use this fact to justify our test for convergence, see section 6. The following properties of tridiagonal matrices are necessary for the discussion below.

Theorem 3. Jennings [25]. (Eigenvalue interlacing.) Let $\nu_1 < \nu_2 < ... < \nu_{m-1}$ denote the eigenvalues of the principal submatrix $B_k$ obtained from $T_m$ by omitting the kth row and column. Then for $1 \leq i \leq m-1$, $\mu_i < \nu_i < \mu_{i+1}$. That is the eigenvalues of $B_k$ interlace the eigenvalues of $T_m$.

In particular, Theorem 3 states that the eigenvalues of $\hat{T}_2$ interlace the eigenvalues of $T_m$.

Theorem 4. Paige [1]. Each eigenvalue $\mu_k$ of $T_m$, $1 \leq k \leq m$, has a unit eigenvector $u^k$ whose ith component satisfies

$$(15) \quad (u_i^k)^2 = a_{i-1}(\mu_k)\hat{a}_{i+1}(\mu_k)/a_m'(\mu_k).$$

The quantities in (15) were defined in section 2. As a Corollary of Theorems 3 and 4 we have

Corollary 1. For any eigenvalue $\mu_k$ of $T_m$,

$$|a_m'(\mu_k)| > \max(|\hat{a}_2(\mu_k)|, |a_{m-1}(\mu_k)|).$$

Theorem 5. In exact arithmetic for any $\mu$ with $a_m(\mu) = 0$, we have

$$(16) \quad \hat{a}_2(\mu)a_{m-1}(\mu) = \prod_{k=2}^{m} \beta_k^2.$$

Proof. In this proof only, we write $a_i$ and $\hat{a}_i$ in place of $a_i(\mu)$ and $\hat{a}_i(\mu)$. Also, we let $\beta_{i,j} = \prod_{k=i}^{j} \beta_k$ with $\beta_{i,i-1} = 1$. By induction on j, $1 \leq i \leq j \leq m$, we prove the more general statement that for any $\mu$ with $a_m(\mu) = 0$,

$$(17) \quad a_{j-1} \hat{a}_{i+1} = \beta_{i+1,j}^2 a_{i-1}\hat{a}_{j+1}.$$

For j=i Equation (17) is trivial. We have the following determinant recursion which can be verified easily by row and column expansions of the determinants.

(18) $\quad a_i \ \hat{a}_{i+1} - \beta_{i+1}^2 a_{i-1} \hat{a}_{i+2} = a_m(\mu) = 0.$

But, (18) is (17) with j=i+1. Assume (17) is valid for $i \leq j \leq k < m$. Then using (17) and the determinant recursions for tridiagonal matrices we have

$$a_k \ \hat{a}_{i+1} = (\mu - \alpha_k) \ \beta_{i+1,k}^2 \ a_{i-1} \ \hat{a}_{k+1} - \beta_k^2 \ \beta_{i+1,k-1}^2 \ a_{i-1} \ \hat{a}_k = \beta_{i+1,k+1}^2 \ a_{i-1} \ \hat{a}_{k+2}.$$

Hence (17) is valid with j=k+1. Equation (16) follows by setting i=1 and j=m in (17). Q.E.D.

In [11] we established an equivalence between the Lanczos tridiagonalization procedure and the conjugate gradient optimization procedure for solving Ax=b, using only local, near-orthogonality of the Lanczos vectors. Because of space limitations we will not list the conjugate gradient recursions and relationships or describe the correspondence between the Lanczos tridiagonalization and conjugate gradients. These can be found in [11]. We note only that the conjugate gradient optimization can be viewed as an iterative procedure. Iterates $x_i$, i=2,... are generated corresponding to residuals $r_i = -Ax_i + b$, which are the errors in solving Ax=b. We have the following Theorem.

Theorem 6. [11]. Let A>0 with distinct eigenvalues $\lambda_k$, $1 \leq k \leq q$. Assume that each of the vectors $v_i$, $1 \leq i \leq m$. generated by the Lanczos recursion satisfy the local near-orthonormality conditions (19) for some $J \geq 1$.

(19) $\quad v_i^T v_j = \varepsilon_{ij}, \ i \neq j, \quad \|v_i\|^2 = 1 + \varepsilon_{ii}, \ 1 \leq |i-j| \leq J.$

We assume that $T_m$ is positive definite and that the norms of the residuals $\rho_i = \|r_i\|$ in the associated conjugate gradient procedure are controlled by

(20) $\quad \rho_j / \rho_i \leq R \quad$ for some R, and $j \geq i$.

Then under suitable conditions on the $\varepsilon$ in (19), on $\lambda_{max} / |\beta|_{min}$, R in (20), and i, we have that the corresponding conjugate gradient directions are locally $\varepsilon - A -$ conjugate, all the conjugate gradient relationships are satisfied approximately, and in fact the norms of the residuals

(21)  $\rho_i = \| r_i \| \to 0$   as $i \to \infty$.

Furthermore, if the starting vector $v_1$ has non-zero projections, $v_1^T z_k \neq 0$ for $1 \leq k \leq q$, for some orthonormal set of eigenvectors of $A$ with $z_k$ corresponding to $\lambda_k$, then for any $\varepsilon > 0$, for large $m$ and all $\lambda_k$, $1 \leq k \leq q$,

(22)  $| a_m(\lambda_k)/a_m | \leq \varepsilon / | v_1^T z_k |$.

Theorem 6 states that (under the relevant hypotheses) for any Lanczos tridiagonalization procedure the associated conjugate gradient procedure converges, and this convergence yields (22).  Equation (22) states that for large enough $m$ every distinct eigenvalue of $A$ is a near-zero of the characteristic equation of $T_m$.  We note that $| a_i(0)/a_i | = 1$.  Thus, the conjugate gradient relationships control the Lanczos process as long as the Lanczos vectors are locally nearly orthonormal.  In practice the size of $m$ required for convergence depends upon the eigenvalue distribution in $A$.  By convergence we mean that each distinct eigenvalue of $A$ is an eigenvalue of $T_m$.  For reasonably uniform distributions $m \leq 2n$, where $n$ is the order of $A$, has proved sufficient.  For matrices with clusters of eigenvalues, at $m = 2n$ approximations to each cluster are obtained.  However, the computation of individual eigenvalues in the clusters may require larger values of $m$.  Being a near zero of $a_m(\mu)$ is of course not the same as being a near root of $a_m(\mu) = 0$.

We have the following pseudo-lemma.  To first order if an eigenvalue $\mu_k$ of $T_m$ is not in a cluster, that is $| a_m'(\mu_k)/a_m | > \delta$, and $\lambda_\ell$ is the eigenvalue of $A$ closest to $\mu_k$ we have

(23)  $| \lambda_\ell - \mu_k | < | a_m(\lambda_\ell)/a_m'(\mu_k) |$.

Hence, $\lambda_\ell$ is a near root of $a_m(\mu) = 0$.  Thus, Theorem 6 gives an explanation or mechanism for the observed convergence of Lanczos eigenvalue procedures that use no reorthogonalization.  The next theorem demonstrates that global orthogonality is not essential.  Well-known relationships that exist for $m < n$ and $V$ orthonormal can also exist when $m > n$ and $V$ arbitrary but of full rank.

Theorem 7. If A has n distinct eigenvalues $\lambda_k$, $1 \leq k \leq n$, and the rank of V is n, then the distinct eigenvalues of the generalized problem

$$(24) \quad V^T A V w = \mu V^T V w$$

are the eigenvalues of A.

Proof. Let $z_k$ be orthonormal eigenvectors of A corresponding to $\lambda_k$. For each k, there is a $u^k$ such that $z_k = V u^k$. Therefore, $V^T A V u^k = \lambda_k V^T V u^k$, so any eigenvalue of A is an eigenvalue of (24). Conversely, if we have $\mu$ and w satisfying (24), set $z = V w$, and (24) yields

$$(25) \quad V^T (A z - \mu z) = 0.$$

But V has full rank.    Q.E.D.

6. Identification Criterion and Estimates of Convergence.    In this section we attempt to justify the test for spurious eigenvalues, and our estimates of the convergence achieved.    As demonstrated in the previous section the conjugate gradient procedure controls the Lanczos tridiagonalization as long as the Lanczos vectors are locally, nearly-orthogonal and the associated residuals do not fluctuate significantly. Since an understanding of the identification criterion requires some knowledge of the convergence estimates, we discuss those estimates first.    We must estimate the accuracy of those eigenvalues of $T_m$ that are approximations to the eigenvalues of A.    Since the Lanczos vectors are not globally nearly-orthogonal, some of the eigenvalues of $T_m$ are spurious.    The identification test described in section 3 identifies the spurious eigenvalues; those remaining are taken as, approximate eigenvalues of A.    For numerically multiple eigenvalues of $T_m$ we have Lemma 1. These eigenvalues are accepted as being accurate to within the tolerance used in determining the multiplicity.

Lemma 1. Let $\mu_i$, $i = k, k+1$, be eigenvalues of $T_m$ with corresponding orthonormal eigenvectors $u^i$, $i = k, k+1$. If $|\mu_k - \mu_{k+1}| < \varepsilon$ then there exists an eigenvalue $\lambda_\ell$ of A with

$$(26) \quad |\mu_k - \lambda_\ell| < \varepsilon' / \|V w\|$$

where $w = c_1 u^k + c_2 u^{k+1}$ with $c_1, c_2$ chosen such that $w_m = 0$ and $\|w\| = 1$. If $z_k = Vu^k$ and $z_{k+1} = Vu^{k+1}$ are independent, then $c_1$ and $c_2$ can be chosen so that $\|Vw\| = 1$.

Proof. Clearly such a $w$ exists and for it from (10) we have $AVw - \mu_k Vw = (\mu_{k+1} - \mu_k)c_2 z_{k+1}$. Using (11) we get (26). Q.E.D.

Therefore, estimates are computed only for the simple, good eigenvalues of $T_m$. Using equations (10), (11), and (14) we see that we can estimate the accuracy of such eigenvalues $\mu$ by estimating the size of $u_m$, the last component of the eigenvector corresponding to $\mu$.

Lemma 2. For any simple eigenvalue $\mu$ of $T_m$ the eigenvector $u$ defined in (15) satisfies

$$(27) \quad |u_m| = |\prod_2^m \beta_j| / [a_m'(\mu)\hat{a}_2(\mu)]^{1/2} \quad \text{and} \quad |u_m/u_1| = |\prod_2^m \beta_j/\hat{a}_2(\mu)|.$$

Proof. By Theorem 4 $(u_m)^2 = a_{m-1}(\mu)/a_m'(\mu)$ and $(u_1)^2 = \hat{a}_2(\mu)/a_m'(\mu)$. Multiplying the numerator and denominator of $|u_m|^2$ and $|u_m/u_1|^2$ by $|\hat{a}_2(\mu)|$ and then using Theorem 5 we obtain (27). Q.E.D.

Lemma 3. [11]. By construction

$$(28) \quad \prod_{j=2}^{m+1} \beta_j = \rho_{m+1} a_m$$

where $\rho_{m+1}$ is the norm of the residual $r_{m+1} = -Ax_{m+1} + b$ generated in the associated conjugate gradient procedure. $a_m$ is the determinant of $T_m$.

Lemma 4. Under the hypotheses of Lemma 2 for each simple $\mu$ and corresponding eigenvector $u$ there is an eigenvalue $\lambda$ of $A$ such that for $z = Vu$,

$$(29) \quad \|z\| \, |\lambda - \mu| \le \rho_{m+1} a_m / [a_m'(\mu)\hat{a}_2(\mu)]^{1/2}.$$

We see from (14), $\| Vu \| > .8$, so that $a'_m(\mu)$ and $\hat{a}_2(\mu)$ determine the size of this bound. $a'_m(\mu)$ is a measure of the density of the eigenvalues near $\mu$. Both also vary according to the location of $\mu$ in the spectrum, near the extreme or in the interior. For interior and extreme eigenvalues with similar neighborhood densities, there is a tendency for the extreme eigenvalues to converge before the interior ones. However, as the example in [23] illustrates interior eigenvalues can converge prior to exterior ones. In example, POIS1008, at $m = 1260$ all of the interior eigenvalues had converged to at least 10 significant digits while 31 percent of the extreme eigenvalues were not converged, 6 percent of which were not even being approximated.

Proof: From (11) there is an eigenvalue $\lambda$ of A such that $\| z \| \, |\lambda - \mu| \leq |\beta_{m+1} u_m|$. From Lemma 3 and (27) we get (29). Q.E.D.

Lemma 2 indicates that for simple eigenvalues of $T_m$, $\hat{T}_2$ is the relevant matrix determining the goodness of $\mu$. Our convergence criterion utilizes either of the two quantities

$$(30) \quad E(\mu) = |\prod_{j=2}^{m+1} \beta_j| / [\hat{a}_2(\mu) a'_m(\mu)]^{1/2} \quad \text{or} \quad E'(\mu) = |\prod_{j=2}^{m+1} \beta_j| / |\hat{a}_2(\mu)|.$$

E is readily computable from the eigenvalues of $\hat{T}_2$ and $T_m$, $E'$ from the eigenvalues of $\hat{T}_2$. We use a subroutine that works separately with the mantissa and exponent of each floating point number. The numerators and denominators in these estimates are computed separately; these quantities can be very small or very large depending upon A. However, the quotients E and $E'$ are properly scaled. Smallness cannot be considered as an absolute measurement, but relative to the other quantities in the expressions.

We note that with $z = Vu$, $E = \| Az - \mu z \|$ and $E' = \| Az - \mu z \| / |u_1|$. Thus $E'$ which requires half the computational effort of E may differ from E by a factor of $10^3$ or more depending upon the size of the first component of the eigenvector u corresponding to $\mu$. However, if the Ritz vector z is an eigenvector of A, then we do not expect $u_1$ to be pathologically small since we have a basic underlying assumption that $v_1^T z$ is not pathologically small.

It is not necessary to compute $E$ or $E'$ for every eigenvalue. Convergence depends upon the relative location in the spectrum, extreme or interior, and the denseness of the neighboring eigenvalues. By sectioning the spectrum, determining the good simple eigenvalue with minimal gap in each section and then computing $E$ for that eigenvalue, we can get a reasonably accurate picture of the overall convergence of all the eigenvalues in that section. If we have a cluster in that section, then typically an eigenvalue in that cluster will have the minimal gap. Minimal gaps are computed using only the good eigenvalues that have been computed up to that point in time. These estimates are, of course, only upper bounds on the true error. However, this type of bound obtained from $\| Az - \mu z \|$ is typically an accurate reflection of the true error and is the type of bound used in most algorithms. We now attempt to justify the identification test. A description of this test was given in section 3 and it is summarized in Table 7.

TABLE 7. Identification Test For Evalues of $T_m$ Using $\hat{T}_2$.

| Case | Eval $\hat{T}_2$ | Multiple Eval of $T_m$ | Conclusion |
|:----:|:----:|:----:|:----:|
| 1 | Yes | Yes | Accept |
| 2 | Yes | No | Reject |
| 3 | No | No | Accept |

For case 1 we have Lemma 1 which tells us that any numerically multiple eigenvalue of $T_m$ is related to $A$. To discuss cases 2 and 3 we introduce $T_{m-1}$, splitting cases 2 and 3 into subcases.

TABLE 8.

| Case | Eval $\hat{T}_2$ | Eval $T_{m-1}$ | Conclusion |
|------|------|------|------|
| 2A | Yes | No | Reject |
| 2B | Yes | Yes | Multiple - Accept |
| 3A | No | Yes | Accept |
| 3B | No | No | Accept, Impossible for Large m |

Our procedure assumes that case 2B implies that $\mu_k$ is multiple and thus case 2B is included in case 1. We do not have a proof of this, just empirical substantiation. For case 2A we have the following Lemma.

<u>Lemma 5</u>. Let $\mu_k$ be a <u>simple eigenvalue</u> of $T_m$. If $\min |\mu_k - \hat{\mu}_j| \leq$ toln, $\min |\mu_k - \bar{\mu}_j| = \delta >>$ toln <u>and</u> gap $\equiv \min\limits_{j \neq k} |\mu_k - \mu_j|$, <u>then</u>

$$(31) \quad |u_m^k / u_1^k|^2 > \delta^2 (gap)/(toln)(\mu_k - \mu_1)(\mu_m - \mu_k), \quad \underline{for}\ k \neq 1, m.$$

Lemma 5 tells us that if $\mu_k$ is an eigenvalue of $\hat{T}_2$ but not of $T_{m-1}$, then $E'(\mu_k)$ is not small. Thus either $u_m^k$ is large or $u_1^k$ is pathologically small and with high probability $\mu_k$ is not an eigenvalue of A. If the V were orthogonal and $z \equiv Vu^k$, then $u_1^k = v_1^T z$ and by assumption this is not small if $\mu_k$ is an eigenvalue of A. If $m > n$, we do not have independence much less orthogonality so that this argument is not directly applicable. However, the numerical results indicate that some modified form of this argument is valid.

<u>Proof</u>. From Theorem 4

$$|u_m^k / u_1^k|^2 = \prod_{j=1}^{m-1} (\mu_k - \bar{\mu}_j) \ / \ \prod_{j=1}^{m-1} (\mu_k - \hat{\mu}_j).$$

By Theorem 3,

$$
(32) \qquad \left. \begin{array}{l} \mu_k - \bar{\mu}_j > \mu_k - \mu_{j+1} \quad 1 \leq j \leq k-2 \\ \bar{\mu}_j - \mu_k > \mu_j - \mu_k \quad k+1 \leq j \leq m-1 \end{array} \right\}
$$

$$
(33) \qquad \left. \begin{array}{l} \mu_k - \hat{\mu}_j < \mu_k - \mu_j \quad 1 \leq j \leq k-1 \\ \hat{\mu}_j - \mu_k < \mu_{j+1} - \mu_k \quad k \leq j \leq m-1 \end{array} \right\}.
$$

Using (32) and (33), w.l.o.g. assuming $\min |\hat{\mu}_j - \mu_k| = \hat{\mu}_k - \mu_k$ since the argument for $\min = \mu_k - \hat{\mu}_{k-1}$ is similar, we get

$$
|u_m^k / u_1^k|^2 > (\mu_{k+1} - \mu_k)(\mu_k - \bar{\mu}_{k-1})(\mu_k - \bar{\mu}_k)/(\mu_k - \mu_1)(\mu_m - \mu_k)(\hat{\mu}_k - \mu_k).
$$

Using the hypotheses we then get (31).  Q.E.D.

For case 3A we have the following lemma, which states that eigenvalues of $T_m$ that are also eigenvalues of $T_{m-1}$ but not of $\hat{T}_2$ are approximate eigenvalues of A.

Lemma 6.  Let $\mu_k$ be a simple eigenvalue of $T_m$. Assume $\min |\mu_k - \bar{\mu}_j| \leq$ toln and $\min |\mu_k - \hat{\mu}_j| = \delta \gg$ toln. Then

$$
(34) \qquad |u_m^k|^2 < \text{toln}/\delta.
$$

Proof.  By Theorem 4

$$
(35) \qquad |u_m^k|^2 = \prod_{j=1}^{m-1} (\mu_k - \bar{\mu}_j) / \prod_{j \neq k}^{m} (\mu_k - \mu_j).
$$

Assume w.l.o.g. that the eigenvalue of $T_{m-1}$ closest to $\mu_k$ is $\bar{\mu}_k$. We have (33) with $\hat{\mu}_j$ replaced by $\bar{\mu}_j$. Using this in (35) and canceling we obtain

$$
|u_m^k|^2 < (\mu_k - \bar{\mu}_k)/(\mu_{k+1} - \mu_k).
$$

But, $\mu_{k+1} - \mu_k > \mu_k - \hat{\mu}_k$ so we obtain (34).  Q.E.D.

From Theorem 5 and Lemma 3, if we have $\rho_m \to 0$ as $m \to \infty$, then case 3B is impossible for large m, in the sense of Theorem 6. For smaller m we accept such eigenvalues as approximations to eigenvalues of A. None of these eigenvalues will have converged. In [8] identification tests based on $T_{m-1}$ were used.

TABLE 9. Identification Test for Evalues of $T_m$ Using $T_{m-1}$

| Case | Eval $T_{m-1}$ | Multiple Eval. of $T_m$ | Conclusion |
|------|----------------|-------------------------|------------|
| 1 | Yes | Yes | Accept |
| 2 | Yes | No | Accept |
| 3 | No | No | Reject |

Tests based on Table 9 will yield reliable results if the tolerance used in cases 1 and 2 is toln in (5). In this case, however, an eigenvalue will not be accepted as an approximation to an eigenvalue of A until it is accurate to toln. Thus, less accurate approximations which occur for fairly small values of m cannot be identified. If the user attempts to increase the size of toln in (5), then the spurious eigenvalues that are converging to copies of converged eigenvalues can be classified incorrectly as good. We note that not all spurious eigenvalues are converging to copies of converged eigenvalues. For an example see [23]. Using $\hat{T}_2$ instead of $T_{m-1}$ we are discarding spurious eigenvalues rather than selecting good ones. Thus, we can obtain approximations to the eigenvalues of A long before these approximations are accurate to toln.

7. Summary. We have presented a Lanczos algorithm with no reorthogonalization for computing many or even all of the eigenvalues of large, sparse real symmetric matrices. It can also be used for computing all the eigenvalues in user-specified subintervals of the spectrum. Convergence in that situation would probably be estimated by computing these eigenvalues for some $T_m$, then incrementing m significantly and recomputing these eigenvalues and confirming that their number

and values have not changed significantly. The arguments for the convergence of this procedure are based upon maintaining local near-orthogonality of the Lanczos vectors and on the relationship of Lanczos tridiagonalization to conjugate gradients. The key to the success of our algorithm is the identification test which selects the subset of the eigenvalues of $T_m$ that do not approximate eigenvalues of A. Estimates of convergence use all of the eigenvalues of $\hat{T}_2$ and more accurate estimates also use all of the eigenvalues of $T_m$. Numerical results demonstrate that this procedure can be very effective.

## REFERENCES

1. C. C. PAIGE, The computation of eigenvalues and eigenvectors of very large sparse matrices, Ph.D. Thesis, University of London (1971).

2. A. K. CLINE, G. H. GOLUB and G. W. PLATZMAN, Calculation of normal modes of oceans using a Lanczos method, Sparse Matrix Computations (ed. J. Bunch and D. Rose), Academic Press, N.Y. (1976).

3. JANE CULLUM and W. E. DONATH, A block, Lanczos algorithm for computing the q algebraically largest eigenvalues and a corresponding eigenspace of large, sparse real symmetric matrices. Proc. 1974 IEEE Conference on Decision and Control, Phoenix, Arizona, (1974), pp. 505-509.

4. JANE CULLUM and W. E. DONATH, A block generalization of the symmetric s-step Lanczos algorithm, RC 4845, IBM Research, Yorktown Heights, N.Y. (1974).

5. J. LEWIS, Algorithms for sparse matrix eigenvalue problems, Technical Report STAN-CS-77-595, Computer Science Dept., Stanford University (1977).

6. M. NEWMAN and A. PIPANO, Fast modal extraction in NASTRAN via the FEER computer program, NASTRAN: User's Experiences, NASA Langley Research Center, Hampton, Virginia (1973), pp. 485-506.

7. C. C. PAIGE, Computational variants of the Lanczos method for the eigenproblem, J. Inst. Math. Appl. 10 (1972), pp. 373-381.

8. J. M. van KATS and H. A. van der VORST, Numerical results of the Paige-style Lanczos method for the computation of extreme eigenvalues of large sparse matrices. Tech. Rpt. TR-3, Academisch Computer Centrum, Utrecht, The Netherlands (1976).

9.  J. M. van KATS and H. A. van der VORST, Automatic monitoring of Lanczos-schemes for symmetric or skew-symmetric generalized eigenvalue problems, Tech. Rpt. TR-7, Academisch Computer Centrum, Utrecht, The Netherlands (1977).

10. T. J. EDWARDS, D. C. LICCIARDELLO and D. J. THOULESS, Use of the Lanczos method for finding complete sets of eigenvalues of large sparse matrices, J. Inst. Math. Appl., (1978), to appear.

11. JANE CULLUM and R. A. WILLOUGHBY, The equivalence of the Lanczos and the conjugate gradient algorithms, RC 6903, IBM Research, Yorktown Heights, N.Y. (1977).

12. JANE CULLUM and R. A. WILLOUGHBY, The Lanczos tridiagonalization and the conjugate gradient algorithms with local $\varepsilon$-orthogonality of the Lanczos vectors, RC 7152, IBM Research, Yorktown Heights, N.Y. (1978).

13. C. C. PAIGE, Error analysis of the Lanczos algorithm for tridiagonalizing a symmetric matrix, J. Inst. Math. Appl., 18, (1976), pp. 341-349.

14. RICHARD UNDERWOOD, An iterative block Lanczos method for the solution of large, sparse symmetric eigenproblems, STAN-CS-75-496, Computer Science Department, Stanford University, (1975). Also Mathematical Software III, ed. J. R. Rice, Academic Press, New York, (1977).

15. AXEL RUHE, Implementation aspects of band Lanczos algorithms for computation of eigenvalues of large sparse symmetric matrices. Tech. Rpt., University of California, San Diego, (1978).

16. T. KAPLAN and L. J. GRAY, Elementary excitations in random substitutional alloys. Phys. Rev. B 14, (1976), pp. 3462-3470.

17. W. KAHAN, Inclusion theorems for clusters of eigenvalues of Hermitian matrices, Dept. of Computer Science, Univ. of Toronto, (1967).

18. W. KAHAN and B. PARLETT, An analysis of Lanczos algorithms for symmetric matrices, Electronics Research Laboratory memorandum, ERL-M467, University of California, Berkeley, (1974).

19. W. KAHAN and B. PARLETT, How far should you go with the Lanczos algorithm? Sparse Matrix Computations, ed. J. Bunch and D. Rose, Academic Press, New York, (1976).

20. B. N. PARLETT and D. S. SCOTT, The Lanczos algorithm with implicit deflation, Electronic Research Laboratory memorandum ERL-M77/70, University of California, Berkeley, (1977).

21. SCOTT KIRKPATRICK and T. P. EGGARTER, Localized states of a binary alloy. Phys. Rev. B 6, (1972), pp. 3598-3609.

22.   EISPAK Guide - Matrix Eigensystem Routines, Lecture Notes in Computer Science, 16, B. T. SMITH et al, 2nd. ed. Springer-Verlag, New York, (1976).

23.   JANE CULLUM and R. A. WILLOUGHBY, Fast modal analysis of large, sparse but unstructured symmetric matrices, RC 7401, IBM Research, Yorktown Heights, N.Y. (1978). Proceedings of the 17th IEEE Conference on Decision and Control, San Diego, Calif., Jan. 10-12, 1979.

24.   J. H. WILKINSON, The Algebraic Eigenvalue Problem, Oxford University Press, London, (1965).

25.   ALAN JENNINGS, Matrix Computation for Engineers and Scientists, Wiley, New York., (1977), p. 281.

# Systolic Arrays (for VLSI)

## H. T. Kung† and Charles E. Leiserson†

*And now I see with eye serene
The very pulse of the machine.*
--William Wordsworth

### Abstract

A systolic system is a network of processors which rhythmically compute and pass data through the system. Physiologists use the word "systole" to refer to the rhythmically recurrent contraction of the heart and arteries which pulses blood through the body. In a systolic computing system, the function of a processor is analogous to that of the heart. Every processor regularly pumps data in and out, each time performing some short computation, so that a regular flow of data is kept up in the network.

Many basic matrix computations can be pipelined elegantly and efficiently on systolic networks having an array structure. As an example, hexagonally connected processors can optimally perform matrix multiplication. Surprisingly, a similar systolic array can compute the LU-decomposition of a matrix. These systolic arrays enjoy simple and regular communication paths, and almost all processors used in the networks are identical. As a result, special purpose hardware devices based on systolic arrays can be built inexpensively using the VLSI technology.

## 1. Introduction

Developments in microelectronics have revolutionized computer design. Integrated circuit technology has increased the number and complexity of components that can fit on a chip or a printed circuit board. Component density has been doubling every one-to-two years and already, a multiplier can fit on a very large scale integrated (VLSI) circuit chip. As a result, the new technology makes it feasible to build low-cost special purpose, peripheral devices to rapidly solve sophisticated problems. Reflecting the changing technology, this paper proposes new multiprocessor

'Authors' address: Department of Computer Science, Carnegie-Mellon University, Pittsburgh, Pennsylvania 15213. This research is supported in part by the National Science Foundation under Grant MCS 75-222-55 and the Office of Naval Research under Contract N00014-76-C-0370, NR 044-422. Charles E. Leiserson is supported in part by the Fannie and John Hertz Foundation.

structures for processing some basic matrix computations.

We are interested in high-performance parallel structures that can be implemented directly as low-cost hardware devices. By performance, we are not refering to the traditional operation counts that characterize classical analyses of algorithms, but rather, the throughput obtainable when a special purpose peripheral device is attached to a general purpose host computer. This implies that time spent in I/O, control, and data movement as well as arithmetic must all be considered. VLSI offers excellent opportunities for inexpensive implementation of high performance devices (Mead and Conway [1978]). Thus, in this paper the cost of a device will be determined by the expense of a VLSI implementation. "Fit the job to the bargain components" -- Blakeslee [1975, p. 4].

VLSI technology has made one thing clear. Simple and regular interconnections lead to cheap implementations and high densities, and high density implies both high performance and low overhead for support components. (Sutherland and Mead [1977] has a good discussion on the importance of having simple and regular geometries for data paths.) For these reasons, we are interested in designing multiprocessor structures which have simple and regular communication paths. We are also interested in employing pipelining as a general method for using these structures. By pipelining, computation may proceed concurrently with input and output, and consequently overall execution time is minimized. Pipelining plus multiprocessing at each stage of a pipeline should lead to the best-possible performance.

Systolic systems provide a realistic model of computation which captures the concepts of pipelining, parallelism and interconnection structures. We do not want to give a formal definition of systolic systems here. For the purpose of this paper, it suffices to view a systolic system as a network of processors which rhythmically compute and pass data through the system. The analogy is to the rhythmic contraction of the heart which pulses blood through the circulatory system of the body. Each processor in a systolic network can be thought of as a heart that pumps multiple streams of data through itself. The regular beating of these parallel processors keeps up a constant flow of data throughout the entire network. As a processor pumps data items through, it performs some constant-time computation and may update some of the items.

Unlike the closed-loop circulatory system of the body, a systolic computing system usually has ports into which inputs flow, and ports where the results of the systolic computation are retrieved. Thus a systolic system can be a pipelined system - input and output occur with every pulsation. This makes them attractive as peripheral processors attached to the data channel of a host computer. Figure 1-1 illustrates how a special purpose systolic device might form a part of a PDP-11 system. A systolic device may also process a real-time data stream or be a component in a larger special purpose system.

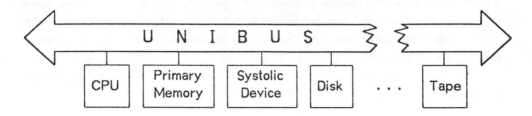

Figure 1-1: A systolic device connected to the UNIBUS of a PDP-11.

This paper deals largely with systolic systems where the underlying network is array structured. (See also Kung and Leiserson [1978].) An array network is attractive for it enjoys simple and regular communication paths. In Section 2, we describe the basic hardware requirements and interconnection schemes for the systolic arrays proposed and discuss the feasibility of building them in VLSI. Section 3 deals with the matrix-vector multiplication problem. Multiplication of two matrices is considered in Section 4. In Section 5, we show that essentially the same systolic arrays for matrix multiplication in Section 4 can be used to find the LU-decomposition of a matrix. Section 6 is concerned with solving triangular linear systems. We show that this problem can be solved by almost the same systolic array for matrix-vector multiplication described in Section 3. Section 7 discusses applications and extensions of the results presented in the previous sections. The applications include the computations of finite impulse response filters, convolutions, and discrete Fourier transforms. Some concluding remarks are given in the last section.

The size of each of our systolic array networks is dependent only on the band width of the band matrix to be processed, and is independent of the length of the band. Thus, a fixed size systolic array can pipeline band matrices with arbitrarily long bands. The pipelining aspect of our arrays is, of course, most effective for band matrices with long bands. Band matrices are interesting in their own right, since many important scientific computations involve band matrices. For these reasons, most of the results in this paper will be presented in terms of their applications to band matrices. All the results apply to dense matrices since a dense matrix can be viewed as a band matrix having the maximum-possible band width.

## 2. The Basic Components and Systolic Array Structures

### 2.1 The Inner Product Step Processor

The single operation common to all the computations considered in this paper is the so-called inner product step, $C \leftarrow C + A \times B$. We postulate a processor which has three registers $R_A$, $R_B$, and $R_C$. Each register has two connections, one for input and one for output. Figure 2-1 shows two types of geometries for this processor.

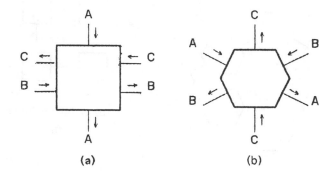

Figure 2-1: Geometries for the inner product step processor.

Type (a) geometry will be used for matrix-vector multiplication and solution of triangular linear systems (Sections 3 and 6), whereas type (b) geometry will be used for matrix multiplication and LU-decomposition (Sections 4 and 5). The processor is capable of performing the inner product step and is called the inner product step processor. We shall define a basic time unit in terms of the operation of this processor. In each unit time interval, the processor shifts the data on its input lines denoted by A, B and C into $R_A$, $R_B$ and $R_C$, respectively, computes $R_C \leftarrow R_C + R_A \times R_B$, and makes the input values for $R_A$ and $R_B$ together with the new value of $R_C$ available as outputs on the output lines denoted by A, B and C, respectively. All outputs are latched and the logic is clocked so that when one processor is connected to another, the changing output of one during a unit time interval will not interfere with the input to another during this time interval. This is not the only processing element we shall make use of, but it will be the work horse. A special processor for performing division will be specified later when it is used.

## 2.2 Systolic Arrays

A systolic device is typically composed of many interconnected inner product step processors. The basic network organization we shall adopt is the mesh-connected scheme in which all connections from a processor are to neighboring processors. (See Figure 2-2.)

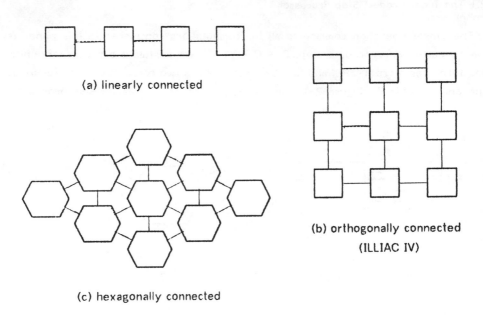

(a) linearly connected

(c) hexagonally connected

(b) orthogonally connected (ILLIAC IV)

Figure 2-2: Mesh-connected systolic arrays.

The most widely known system based on this organization is the ILLIAC IV (Barnes et al. [1968]). If diagonal connections are added in one direction only, we shall call the resulting scheme hexagonally mesh-connected or hex-connected for short. We shall demonstrate that linearly connected and hex-connected arrays are natural for matrix problems.

Processors lying on the boundary of the systolic array may have external connections to the host memory. Thus, an input/output data path of a boundary processor may sometimes be designated as an external input/output connection for the device. A boundary processor may receive input from the host memory through such an external connection, or it may receive a fixed value such as zero. On the other hand, a boundary processor can send data to the host memory through an external output connection. An output of a boundary processor may sometimes be ignored. This will be designated by omitting the corresponding output line.

In this paper we assume that the processors in a systolic array are synchronous as described in Section 2.1. However, it is possible to view the processors being asynchronous, each computing its output values when all its inputs are available, as in a data flow model. For the results of this paper we believe the synchronous approach to be more direct and intuitive.

The hardware demands of the systolic arrays in this paper are readily seen to be modest. The processing elements are uniform, interprocessor connections are simple and regular, and external connections are minimized. It is our belief that construction of these systolic arrays will prove to be cost-effective using, for instance, the modern VLSI technology.

## 3. Matrix-Vector Multiplication on a Linear Systolic Array

We consider the problem of multiplying a matrix $A = (a_{ij})$ with a vector $x = (x_1,...,x_n)^T$. The elements in the product $y = (y_1,...,y_n)^T$ can be computed by the following recurrences.

$$y_i^{(1)} = 0,$$

$$y_i^{(k+1)} = y_i^{(k)} + a_{ik}x_k,$$

$$y_i = y_i^{(n+1)}.$$

Suppose $A$ is an nxn band matrix with band width $w = p+q-1$. (See Figure 3-1 for the case when $p = 2$ and $q = 3$.) Then the above recurrences can be evaluated by pipelining the $x_i$ and $y_i$ through a systolic array consisting of $w$ linearly connected inner product step processors. We illustrate the operation of the systolic array for the band matrix-vector multiplication problem in Figure 3-1. For this case the linearly connected systolic array has four inner product step processors. See Figure 3-2.

The general scheme of the computation can be viewed as follows. The $y_i$, which are initially zero, are pumped to the left while the $x_i$ are pumped to the right and the $a_{ij}$ are marching down. (For the general problem of computing $Ax+d$ where $d=(d_1,...,d_n)^T$ is any given vector, $y_i$ should be initialized as $d_i$). All the moves are synchronized. It turns out that each $y_i$ is able to accumulate all its terms, namely, $a_{i,i-2}x_{i-2}$, $a_{i,i-1}x_{i-1}$, $a_{i,i}x_i$ and $a_{i,i+1}x_{i+1}$, before it leaves the network. Figure 3-3 illustrates the first seven pulsations of the systolic array. Note that when $y_1$ and $y_2$ are output they have the correct values. Observe also that at any given time

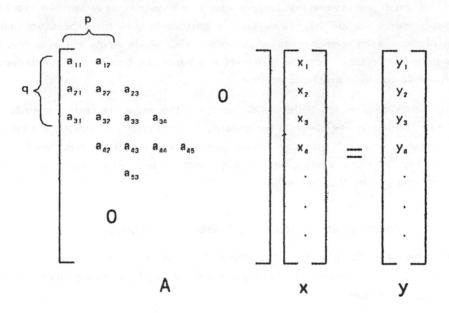

Figure 3-1: Multiplication of a vector by a band matrix with p = 2 and q = 3.

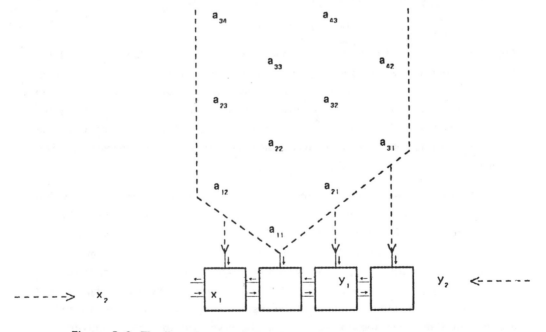

Figure 3-2: The linearly connected systolic array for the matrix-vector multiplication problem in Figure 3-1.

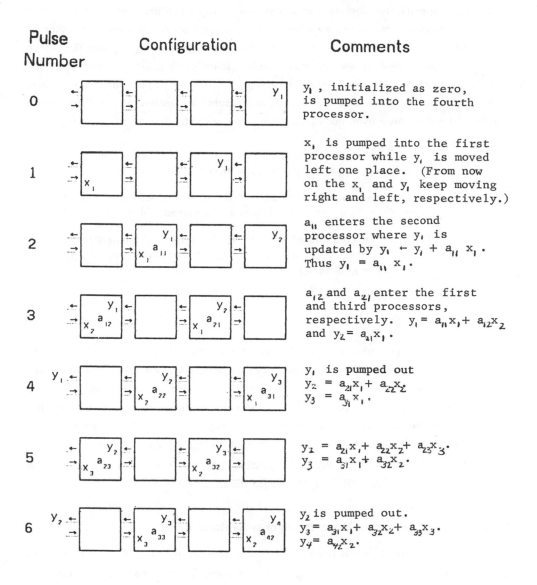

| Pulse Number | Configuration | Comments |
|---|---|---|
| 0 | | $y_1$, initialized as zero, is pumped into the fourth processor. |
| 1 | | $x_1$ is pumped into the first processor while $y_1$ is moved left one place. (From now on the $x_i$ and $y_i$ keep moving right and left, respectively.) |
| 2 | | $a_{11}$ enters the second processor where $y_1$ is updated by $y_1 \leftarrow y_1 + a_{11} x_1$. Thus $y_1 = a_{11} x_1$. |
| 3 | | $a_{12}$ and $a_{21}$ enter the first and third processors, respectively. $y_1 = a_{11}x_1 + a_{12}x_2$ and $y_2 = a_{21}x_1$. |
| 4 | | $y_1$ is pumped out $y_2 = a_{21}x_1 + a_{22}x_2$ $y_3 = a_{31}x_1$. |
| 5 | | $y_2 = a_{21}x_1 + a_{22}x_2 + a_{23}x_3$. $y_3 = a_{31}x_1 + a_{32}x_2$. |
| 6 | | $y_2$ is pumped out. $y_3 = a_{31}x_1 + a_{32}x_2 + a_{33}x_3$. $y_4 = a_{42}x_2$. |

Figure 3-3: The first seven pulsations of the linear systolic array in Figure 3-2.

alternate processors are idle. Indeed, by coalescing pairs of adjacent processors, it is possible to use w/2 processors in the network for a general band matrix with band width w.

We now specify the operation of the systolic array more precisely. Assume that the processors are numbered by integers 1, 2, . . ., w from the left end processor to the right end processor. Each processor has three registers, $R_A$, $R_x$ and $R_y$, which will hold entries in A, x and y, respectively. Initially, all registers contain zeros. Each pulsation of the systolic array consists of the following operations, but for odd numbered pulses only odd numbered processors are activated and for even numbered pulses only even numbered processors are activated.

1. *Shift.*

  - $R_A$ gets a new element in the band of matrix A.

  - $R_x$ gets the contents of register $R_x$ from the left neighboring node. (The $R_x$ in processor 1 gets a new component of x.)

  - $R_y$ gets the contents of register $R_y$ from the right neighboring node. (Processor 1 outputs its $R_y$ contents and the $R_y$ in processor w gets zero.)

2. *Multiply and Add.*

$$R_y \leftarrow R_y + R_A \times R_x.$$

Using the type (a) inner product step processor postulated in section 2, we note that the three shift operations in step 1 can be done simultaneously, and that each pulsation of the systolic array takes a unit of time. Suppose the bandwidth of A is w = p+q-1. It is readily seen that after w units of time the components of the product y = Ax are pumped out from the left end processor at the rate of one output every two units of time. Therefore, using our systolic network all the n components of y can be computed in 2n+w time units, as compared to the O(wn) time needed for a sequential algorithm on a uniprocessor computer.

## 4. Matrix Multiplication on a Hexagonal Systolic Array

This section considers the problem of multiplying two nxn matrices. It is easy to see that the matrix product C = $(c_{ij})$ of A = $(a_{ij})$ and B = $(b_{ij})$ can be computed by the following recurrences.

$$c_{ij}^{(1)} = 0,$$

$$c_{ij}^{(k+1)} = c_{ij}^{(k)} + a_{ik}b_{kj},$$

$$c_{ij} = c_{ij}^{(n+1)}.$$

Let A and B be nxn band matrices of band width $w_1$ and $w_2$, respectively. We show how the recurrences above can be evaluated by pipelining the $a_{ij}$, $b_{ij}$ and $c_{ij}$ through a systolic array having $w_1 w_2$ hex-connected inner product step processors. We illustrate the general scheme by considering the matrix multiplication problem depicted in Figure 4-1. The diamond shaped systolic array for this case is shown in Figure 4-2, where processors are hex-connected and data flows are indicated by arrows.

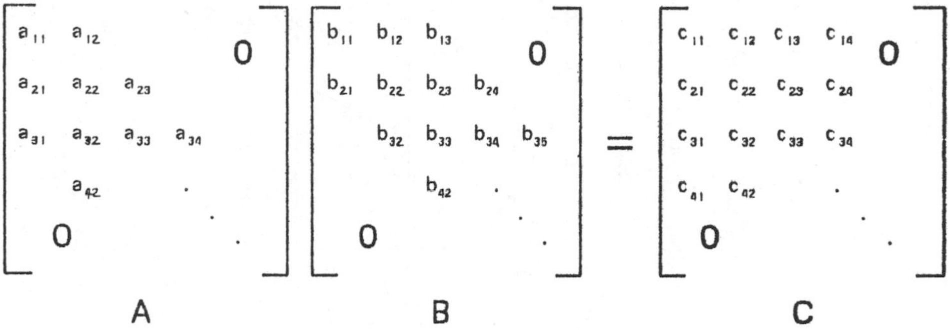

$$
\begin{bmatrix}
a_{11} & a_{12} & & & & \\
a_{21} & a_{22} & a_{23} & & 0 & \\
a_{31} & a_{32} & a_{33} & a_{34} & & \\
& a_{42} & & & \ddots & \\
0 & & & & & \ddots
\end{bmatrix}
\begin{bmatrix}
b_{11} & b_{12} & b_{13} & & & \\
b_{21} & b_{22} & b_{23} & b_{24} & 0 & \\
& b_{32} & b_{33} & b_{34} & b_{35} & \\
& & b_{42} & & \ddots & \\
0 & & & & & \ddots
\end{bmatrix}
=
\begin{bmatrix}
c_{11} & c_{12} & c_{13} & c_{14} & & \\
c_{21} & c_{22} & c_{23} & c_{24} & 0 & \\
c_{31} & c_{32} & c_{33} & c_{34} & & \\
c_{41} & c_{42} & & & \ddots & \\
0 & & & & & \ddots
\end{bmatrix}
$$

$$\text{A} \qquad\qquad\qquad \text{B} \qquad\qquad\qquad \text{C}$$

Figure 4-1: Band matrix multiplication.

The elements in the bands of A, B and C are pumped through the systolic network in three directions synchronously. Each $c_{ij}$ is initialized to zero as it enters the network through the bottom boundaries. (For the general problem of computing AB+D where $D=(d_{ij})$ is any given matrix, $c_{ij}$ should be initialized as $d_{ij}$.) One can easily see that with the type (b) inner product step processors described in Section 2, each $c_{ij}$ is able to accumulate all its terms before it leaves the network through the upper boundaries. Figure 4-3 shows four consecutive pulsations of the hexagonal systolic array. The reader is invited to study the data flow of this problem more closely by making transparencies of the band matrices shown in the figures, and moving them over the network picture as described above.

   Let A and B be nxn band matrices of band width $w_1$ and $w_2$, respectively. Then a systolic array of $w_1 w_2$ hex-connected processors can pipeline the matrix multiplication AxB in $3n+\min(w_1, w_2)$ units of time. Note that in any row or column of the network, out of every three consecutive processors, only one is active at given time. It is possible to use about $(w_1 w_2)/3$ processors in the network for multiplying two band matrices with band widths $w_1$ and $w_2$.

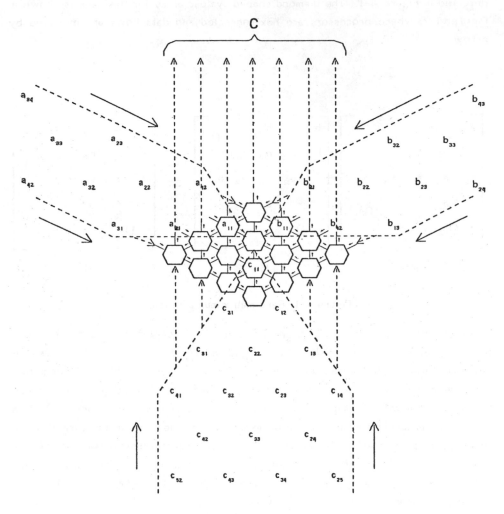

Figure 4-2: The hex-connected systolic array for the matrix multiplication problem in Figure 4-1.

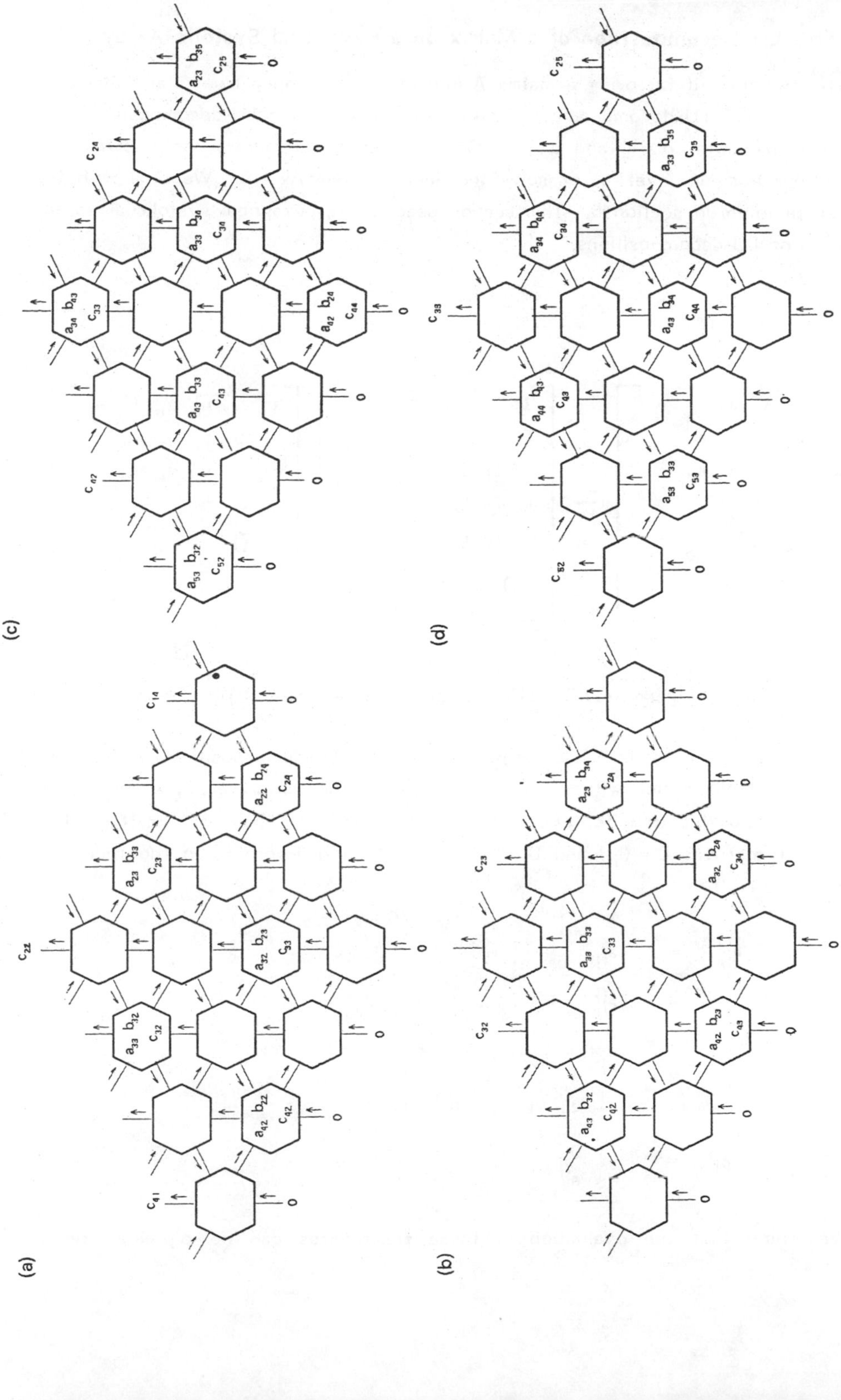

Figure 4-3: Four pulsations of the hexagonal systolic array in Figure 4-2.

## 5. The LU-Decomposition of a Matrix on a Hexagonal Systolic Array

The problem of factoring a matrix A into lower and upper triangular matrices L and U is called LU-decomposition. Figure 5-1 illustrates the LU-decomposition of a band matrix with $p = 4$ and $q = 4$. Once the L and U factors are known, it is relatively easy to invert A or solve the linear system $Ax = b$. We deal with the latter problem in section 6. This section describes a hexagonal systolic array for computing LU-decompositions.

$$
\begin{bmatrix}
a_{11} & a_{12} & a_{13} & a_{14} & & & & \mathbf{0} \\
a_{21} & a_{22} & a_{23} & a_{24} & a_{25} & & & \\
a_{31} & a_{32} & a_{33} & a_{34} & a_{35} & & & \\
a_{41} & a_{42} & a_{43} & & & & & \\
& a_{52} & a_{53} & & & & & \\
\mathbf{0} & & & & & & & 
\end{bmatrix}
=
\begin{bmatrix}
1 & & & & & & \mathbf{0} \\
l_{21} & 1 & & & & & \\
l_{31} & l_{32} & 1 & & & & \\
l_{41} & l_{42} & l_{43} & 1 & & & \\
& l_{52} & l_{53} & & & & \\
\mathbf{0} & & & & & & \ddots
\end{bmatrix}
\cdot
\begin{bmatrix}
u_{11} & u_{12} & u_{13} & u_{14} & & & \mathbf{0} \\
& u_{22} & u_{23} & u_{24} & u_{25} & & \\
& & u_{33} & u_{34} & u_{35} & & \\
& & & & & & \\
\mathbf{0} & & & & & & \ddots
\end{bmatrix}
$$

$$\qquad\qquad A \qquad\qquad\qquad\qquad L \qquad\qquad\qquad\qquad U$$

**Figure 5-1: The LU-decomposition of a band matrix.**

We assume that matrix A has the property that its LU-decomposition can be done by Gaussian elimination without pivoting. (This is true, for example, when A is a symmetric positive-definite, or an irreducible, diagonally dominant matrix.) The triangular matrices $L = (l_{ij})$ and $U = (u_{ij})$ are evaluated according to the following recurrences.

$$a_{ij}^{(1)} = a_{ij},$$

$$a_{ij}^{(k+1)} = a_{ij}^{(k)} + l_{ik}(-u_{kj}),$$

$$l_{ik} = \begin{cases} 0 & \text{if } i < k, \\ 1 & \text{if } i = k, \\ a_{ik}^{(k)} u_{kk}^{-1} & \text{if } i > k, \end{cases}$$

$$u_{kj} = \begin{cases} 0 & \text{if } k > j, \\ a_{kj}^{(k)} & \text{if } k \le j. \end{cases}$$

We show that the evaluation of these recurrences can be pipelined on a hex-connected systolic array of hex-connected processors. A global view of this pipelined computation is shown in Figure 5-2 for the LU-decomposition problem depicted in Figure 5-1. The systolic array in Figure 5-2 is constructed as follows. The processors below the upper boundaries are the standard type (b) inner product step processors and are hex-connected exactly same as the matrix multiplication network presented in Section 4. The processor at the top, denoted by a circle, is a special processor. It computes the reciprocal of its input and pumps the result southwest, and also pumps the same input northward unchanged. The other processors on the upper boundaries are again type (b) inner product step processors, but their orientation is changed: the ones on the upper left boundary are rotated 120 degrees clockwise; the ones on the upper right boundary are rotated 120 degrees counterclockwise.

The flow of data on the systolic array is indicated by arrows in the figure. As in the hexagonal systolic array for matrix multiplication , each processor only operates every third time pulse. Figure 5-3 illustrates four consecutive pulsations of the systolic array. Note that in the figure, because A is a band matrix with p = 4 and q = 4 we have that $a_{i+3,i}^{(k)} = a_{i+3,i}$ and $a_{i,i+3}^{(k)} = a_{i,i+3}$ for $1 \leq k \leq i$ and $i \geq 2$. Thus $a_{52}$, for example, can be viewed as $a_{52}^{(2)}$ when it enters the network.

There are several equivalent systolic arrays that reflect only minor changes to the network presented in this section. For example, the elements of L and U can be retrieved as output in a number of different ways. Also, the "-1" input to the network can be changed to a "+1" if the special processor at the top of the network computes minus the reciprocal of its input.

If A is an nxn band matrix with band width w = p+q-1, a systolic array having no more than pq hex-connected processors can compute the LU-decomposition of A in 3n+min(p,q) units of time. If A is an nxn dense matrix, this means that $n^2$ hex-connected processors can compute the L and U matrices in 4n units of time which includes I/O time.

The remarkable fact that the matrix multiplication network forms a major part of the LU-decomposition network is due to the similarity of their defining recurrences. In any row or column of the LU-decomposition systolic array, only one out of every three consecutive processors is active at a given time. As we observed for matrix multiplication, the number of processors can be reduced to about pq/3.

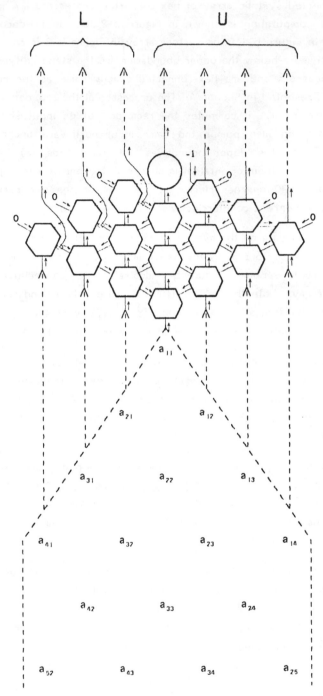

Figure 5-2: The hex-connected systolic array for pipelining the LU-decomposition of the band matrix in Figure 5-1.

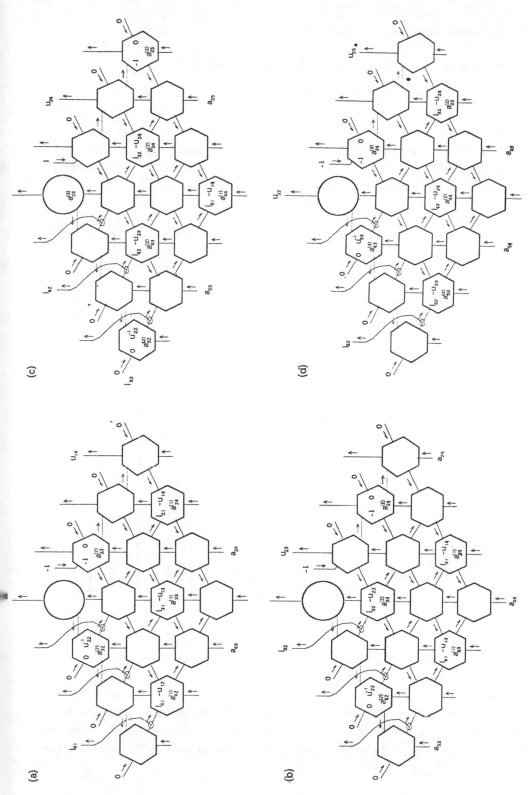

Figure 5-3: Four pulsations of the hexagonal systolic array in Figure 5-2.

## 6. Solving a Triangular Linear System on a Linear Systolic Array

Suppose that we want to solve a linear system $Ax = b$. Then after having done the LU-decomposition of A (e.g., by methods described in Section 5), we still have to solve two triangular linear systems $Ly = b$ and $Ux = y$. This section concerns itself with the solution of triangular linear systems. An upper triangular linear system can always be rewritten as a lower triangular linear system. Without loss of generality, this section deals exclusively with lower triangular linear systems.

Let $A = (a_{ij})$ be a nonsingular nxn band lower triangular matrix. Suppose that A and an n-vector $b = (b_1,...,b_n)^T$ are given. The problem is to compute $x = (x_1,...,x_n)^T$ such that $Ax = b$. The vector x can be computed by forward substitution:

$$y_i^{(1)} = 0,$$

$$y_i^{(k+1)} = y_i^{(k)} + a_{ik}x_k,$$

$$x_i = (b_i - y_i^{(i)})/a_{ii}.$$

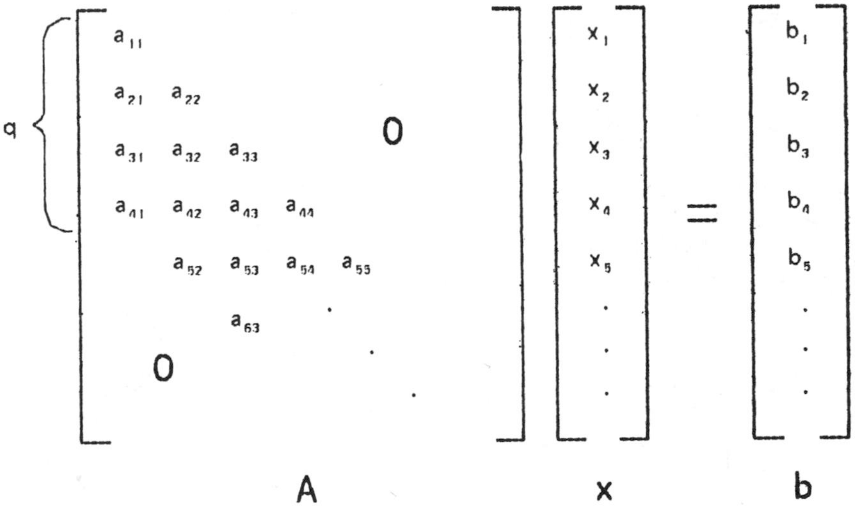

Figure 6-1: The band (lower) triangular linear system where q = 4.

Suppose that A is a band matrix with band width $w = q$. (See Figure 6-1 for the case when q = 4.) Then the above recurrences can be evaluated by a systolic array similar to that used for band matrix-vector multiplication in Section 3. (Observe the similarity of the defining recurrences for these two problems.) We illustrate our result by considering the linear system problem in Figure 6-1. For this case, the systolic array is described in Figure 6-2.

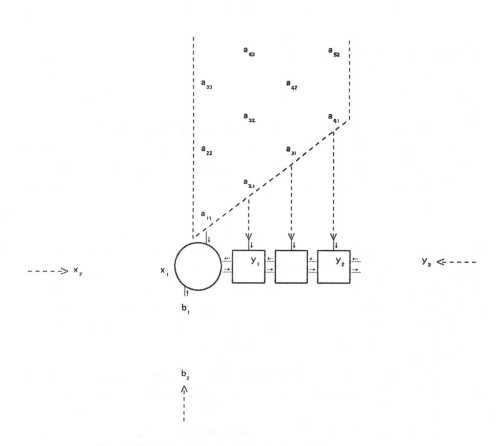

**Figure 6-2:** The linearly connected systolic array for solving the triangular linear system in Figure 6-1.

The $y_i$, which are initially zero, are forced leftward through the systolic array while the $x_i$, $a_{ij}$ and $b_i$ are pumped as indicated in Figure 6-2. The left end processor is special in that it performs $x_i \leftarrow (b_i - y_i)/a_{ii}$. (In fact, the special processor introduced in section 5 to solve the LU-decomposition problem is a special case of this more general processor.) Each $y_i$ accumulates inner product terms in the rest of the processors as it moves to the left. At the time $y_i$ reaches the left end processor it has the value $a_{i1}x_1 + a_{i2}x_2 + \ldots + a_{i,i-1}x_{i-1}$, and, consequently, the $x_i$ computed by $x_i \leftarrow (b_i - y_i)/a_{ii}$ at the processor will have the correct value. Figure 6-3 demonstrates the first seven pulsations of the systolic array. From the figure one can check that the final values of $x_1$, $x_2$, $x_3$ and $x_4$ are all correct. With this systolic array we can solve an nxn band triangular linear system with band width w = q in 2n+q units of time. As we observed for the matrix-vector multiplication problem, the number of processors required by the array can be reduced to w/2.

## 7. Applications and Comments

### 7.1 Variants of the Systolic Array

If more information is available about the specific matrices involved, an optimized version of the systolic arrays presented above can be used. It is important that the reader understands the basic principles so that he can construct appropriate variants for his specific problems. No attempt is made here to list all the possible variants.

As pointed out in Section 1, although most of our illustrations are of band matrices, all the systolic arrays work for regular nxn dense matrices. In this case the band width of the matrix is w = 2n-1. If the band width of a matrix is so large that it requires more processors than a given array provides, then one should decompose the matrix and solve each subproblem on the network. Thus, for example, the matrix multiplication of two nxn matrices or the LU-decomposition of an nxn matrix can be done in $O(n^3/k^2)$ time on a kxk systolic array.

One can often reduce the number of processors required by a systolic array if the matrix is known to be sparse or symmetric. For example, the matrices arising from a set of finite differences or finite elements approximations to differential equations are usually "sparse band matrices". These are band matrices whose nonzero entries appear only in a few of those lines in the band which are parallel to the diagonal. In this case by introducing proper delays to each processor for shifting its data to its neighbors, the number of processors required by the systolic array in Section 3 can be reduced to the number of those diagonals which contain nonzero entries. This variant is useful for performing iterative methods involving sparse band matrices. Another example concerns the LU-decomposition problem considered in Section 5. If matrix A is symmetric positive definite, then it is possible to use only the left portion of the hex-connected network, since in this case U is simply $DL^T$ where D is the

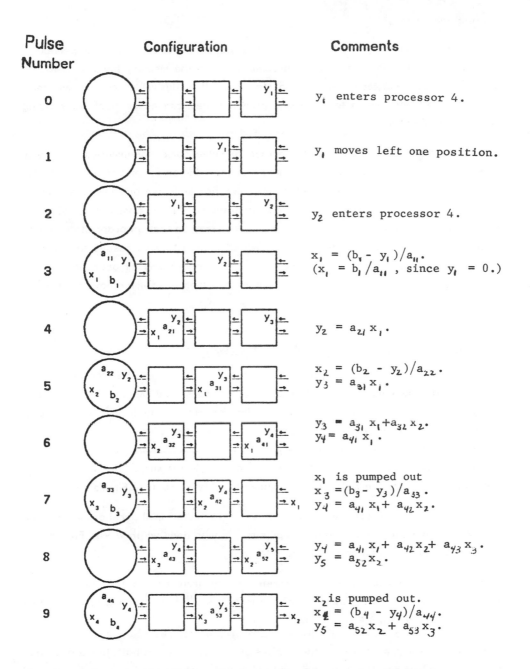

Figure 6-3: Solving a lower band triangular system.

diagonal matrix $(a_{kk}^{(k)})$.

The optimal choice of the size of the systolic network to solve a particular problem depends upon not only the problem but also the memory bandwidth to the host computer. For achieving high performance, it is desirable to have as many processors as possible in the network, provided they can all be kept busy doing useful computations.

It is possible to use our systolic arrays to solve some nonnumerical problems when appropriate interpretations are given to the addition (+) and multiplication (x) operations. For example, some pattern matching problems can be viewed as matrix problems with comparison and Boolean operations. It is possible to store a dynamically changing data structure in a systolic array so that an order statistic can always be determined in constant time. We shall report these results in a future paper. It can be instructive to view the + and x operations as operations in an abstract algebraic structure such as a semiring and then to examine how our results hold in such an abstract setting.

## 7.2 Convolution, Filter, and Discrete Fourier Transform

There are a number of important problems which can be formulated as matrix-vector multiplication problems and thus can be solved rapidly by the systolic array in Section 3. The problems of computing convolutions, finite impulse response (FIR) filters, and discrete Fourier transforms are such examples. If a matrix has the property that the entries on any line parallel to the diagonal are all the same, then the matrix is a Toeplitz matrix. The convolution problem is simply the matrix-vector multiplication where the matrix is a triangular Toeplitz matrix (see Figure 7-1).

A p-tap FIR filter can be viewed as a matrix-vector multiplication where the matrix is a band upper triangular Toeplitz matrix with band width w = p. Figure 7-2 represents the computation of a 4-tap filter.

On the other hand, an n-point discrete Fourier transform is the matrix-vector multiplication, where the (i,j) entry of the matrix is $\omega^{(i-1)(j-1)}$ and $\omega$ is a primitive $n^{th}$ root of unity. (See Figure 7-3).

Therefore using a linearly connected systolic array of size n both the convolution of two n-vectors and the n-point discrete Fourier transform can be computed in O(n) units of time, rather than O(n log n) as required by the sequential FFT algorithm. Moreover, note that for the convolution and filter problems each processor has to receive an entry of the matrix only once, and this entry can be shipped to the

$$\begin{bmatrix} a_1 & & & & \\ a_2 & a_1 & & & \\ a_3 & a_2 & a_1 & & \quad\quad 0 \\ a_4 & a_3 & a_2 & a_1 & \\ a_5 & a_4 & a_3 & a_2 & a_1 \\ & & & \\ & \cdot & \cdot & \cdot \end{bmatrix} \begin{bmatrix} x_1 \\ x_2 \\ x_3 \\ x_4 \\ x_5 \\ \cdot \\ \cdot \\ \cdot \end{bmatrix} = \begin{bmatrix} b_1 \\ b_2 \\ b_3 \\ b_4 \\ b_5 \\ \cdot \\ \cdot \\ \cdot \end{bmatrix}$$

Figure 7-1: The convolution of vectors a and x.

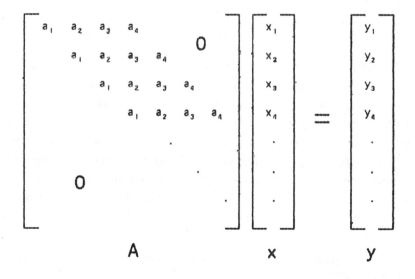

$$\begin{matrix} A & x & y \end{matrix}$$

Figure 7-2: A 4-tap FIR filter with coefficients $a_1$, $a_2$, $a_3$, and $a_4$.

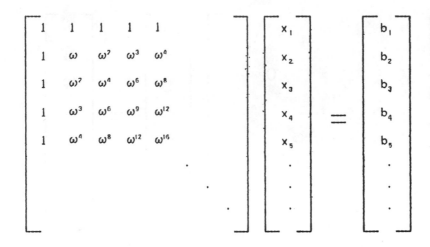

Figure 7-3: The discrete Fourier transform of vector x.

processor through horizontal connections and stay in the processor during the rest of the computation. For the discrete Fourier transform problem each processor can in fact generate on-the-fly the powers of $\omega$ it requires. As a result, for these three problems it is not necessary for each processor in the network to have the external input connection on the top of the processor, as depicted in Figure 3-2.

In the following we describe how the powers of $\omega$ can be generated on-the-fly during the process of computing an n-point discrete Fourier transform. The requirement is that if a processor is i units apart from the middle processor then at time i + 2j the processor must have the value of $\omega^{j^2 + ij}$ for all i, j. This requirement can be fulfilled by using the algorithm below. We assume that each processor has one additional register $R_t$. All processors except the middle one perform the following operations in each step, but for odd (respectively, even) numbered time steps only processors which are odd (even) units apart from the middle processor are activated. For all processors except the middle one the contents of both $R_A$ and $R_t$ are initially zero.

1. *Shift.* If the processor is in the left (respectively, right) hand side of the middle processor then

    - $R_A$ gets the contents of register $R_A$ from the right (respectively, left) neighboring processor.

    - $R_t$ gets the contents of register $R_t$ from the right (respectively, left) neighboring processor.

2. *Multiply.*

$$R_A \leftarrow R_A \times R_t.$$

The middle processor is special; it performs the following operations at every ever numbered time step. For this processor the contents of both $R_A$ and $R_t$ are initially one.

1. $R_A \leftarrow R_A \times R_t^2 \times \omega.$

2. $R_t \leftarrow R_t \times \omega.$

## 7.3 The Common Memory Access Pattern

Note that all the systolic arrays given in this paper store and retrieve elements of the matrix in the same order.. (See Figures 3-2, 4-2, 5-2, and 6-2.) Therefore, we recommend that matrices be always arranged in memory according to this particular ordering so that they can be accessed efficiently by any of the systolic structures.

## 7.4 The Pivoting Problem, and Orthogonal Factorization

In section 5 we assume that the matrix A has the property that there is no need of using pivoting when Gaussian elimination is applied to A. What should one do if A does not have this nice property? (Note that Gaussian elimination becomes very inefficient on mesh-connect processors if pivoting is necessary.) This question motivated us to consider Givens' transformation (see, for example, Hammering [1974]) for triangularizing a matrix, which is known to be a numerically stable method. It turns out that, like Gaussian elimination without pivoting, the orthogonal factorization based on Givens' transformation can be implemented naturally on mesh-connected processors, although a pipelined systolic array implementation appears to be more complex. Our results on Givens' transformation will be reported in another paper. (Sameh and Kuck [1978] considers parallel linear system solvers based on Givens' transformation, but they do not give solutions to the processor communication problem considered in this paper.)

# 8. Concluding Remarks

Systolic structures provide a model of computation for studying parallel algorithms for VLSI. The model takes into account issues such as I/O, control, and interprocessor communication. In a systolic system pipelining can overlap I/O with computation to ensure high throughput. Since loading of data into the network occurs naturally as computation proceeds, no extra control logic is required. Nor is initialization logic needed. Communication among processors is through fixed data paths. For a low cost and high performance implementation in VLSI (or even printed circuit technology), it is desireable that these paths have simple and regular geometries. These reasons make systolic arrays considered in this paper especially attractive. Indeed, interconnection structures other than arrays exist which satisfy these constraints. Future work will examine some of these connection schemes and demonstrate that systolic systems generalize beyond simple cellular structures.

We have discovered that some data flow patterns are fundamental in matrix computations. For example, the two-way flow on the linearly connected network is common to both matrix-vector multiplication and solution of triangular linear systems (Sections 3 and 6), and the three-way flow on the hexagonally mesh-connected network is common to both matrix multiplication and LU-decomposition (Sections 4 and 5). A practical implication of this fact is that one systolic device may be used for solving many different problems. Moreover, we note that almost all the processors needed in any of these devices are the inner product step processor postulated in Section 2. A careful design of this processor is desirable since it is the work horse for all the devices presented.

Research in interconnection networks and algorithms has been frequently motivated by parallel array computers such as ILLIAC IV. (See, for example, Kuck [1968, 1977] and Stone [1975].) Although the results presented in this paper were motivated by the advance in VLSI, they reach beyond. The systolic arrays in this paper can be implemented as efficient algorithms on traditional parallel array machines.

For the important problem of solving a dense system of n linear equations in $O(n)$ time on $n^2$ mesh-connected processors, we have improved upon the recent results of Kant and Kimura [1978]. The basis of their results is a theorem on determinants which was known to J. Sylvester in 1851. Their algorithm requires that the matrix be "strongly nonsingular" in the sense that every square submatrix is nonsingular. It is sufficient for our algorithms that the matrix be symmetric positive-definite or irreducible diagonally dominant.

Hoare [1977], and Thurber and Wald [1975] describe some matrix multiplication algorithms on an orthogonally connected processor array. Unlike our results, their algorithms require that one or more of the three matrices involved in matrix multiplication stay in the array statically during the computation. This introduces overheads in I/O time and control logic for loading the array with the static matrix. Our systolic array makes use of the hexagonal connections to pipeline all three matrices.

Processor communication will likely dominate the cost of parallel algorithms and systems. Communication paths inherently require more space and energy than processing elements do. We regard the problem of minimizing communication costs as fundamental, and we believe systolic structures provide models that can bridge the gap between theory and practice. Systolic arrays can be built in VLSI. Connected to a standard Von Neumann computer, a systolic device provides inexpensive but massive computation power.

## REFERENCES

**Barnes et al.** [1968] Barnes, G. H., Brown, R. M., Maso, K., Kuck, D. J., Slotnick, D. L., and Stokes, R. A., "The ILLIAC IV Computer," IEEE Transactions on Computers C-17 (1968), pp. 746-757.

**Blakeslee** [1975] Blakeslee, Thomas R., Digital Design with Standard MSI and LSI, John Wiley & Sons, New York, 1975.

**Hammering** [1974] Hammering, S., "A Note on Modifications to the Givens' Plane Rotation," J. Inst. Math. Appl. 13 (1974), pp. 215-218.

**Hoare** [1977] Hoare, C. A. R., "Communicating Sequential Processes," Communications of the ACM 21 (1978), pp. 666-677.

**Kant and Kimura** [1978] Kant, Rajani M. and Kimura, Takayuki, "Decentralized Parallel Algorithms for Matrix Computation," Proceedings of the Fifth Annual Symposium on Computer Architecture, Palo Alto, California, April 1978, pp. 96-100.

**Kuck** [1968] Kuck, D. J., "ILLIAC IV Software and Application Programming," IEEE Transactions on Computers C-17 (1968), pp. 758-770.

**Kuck** [1977] Kuck, D. J., "A Survey of Parallel Machine Organization and Programming," Computing Surveys 9 (1977), pp. 29-59.

**Kung and Leiserson** [1978] Kung, H. T. and Leiserson, Charles E., "Systolic Array Apparatuses for Matrix Computations," U. S. Patent Application, Filed December 1978.

**Mead and Conway** [1978] Mead, C. A. and Conway, L. A., Introduction to VLSI Systems, 1978.

**Sameh and Kuck** [1978] Sameh, A. H. and Kuck, D. J., "On Stable Parallel Linear System Solvers," Journal of the Association for Computing Machinery 25 (1978), pp. 81-91.

Stone [1975] Stone, Harold S., "Parallel Computations", Introduction to Computer Architecture, edited by H. S. Stone, Science Research Associates, Chicago, 1975, pp. 318-374.

Sutherland and Mead [1977] Sutherland, Ivan E. and Mead, Carver A., "Microelectronics and Computer Science," Scientific American 237 (1977), pp. 210-228.

Thurber and Wald [1975] Thurber, Kenneth J., and Wald, Leon D., "Associative and Parallel Processors," Computing Surveys 7 (1975), pp. 215-255.

# A Basis Factorization Method
# for Block Triangular Linear Programs[1]

## André F. Perold[2] and George B. Dantzig

ABSTRACT: Time-staged and multi-staged linear programs usually have a structure that is block triangular. Basic solutions to such problems typically have the property that similar type activities persist in the basis over several consecutive time-periods. When this occurs the basis is close to being square block triangular. In 1955 Dantzig suggested a way of factorizing the basis to take advantage of this property. This paper discusses persistence in staircase models and then presents a considerable refinement of Dantzig's factorization algorithm. The method has been implemented in an experimental code, with use being made of LU and QR factorization and updating techniques for the solution of small sub-systems of equations. Computational experience on several dynamic models is reported.

## 1. Introduction

This paper presents a basis factorization method for linear programs with a large sparse block triangular structure. Such linear programs form an important class of models, including all multi-staged and time-staged linear programs that arise either naturally as, for example, dynamic economic models of investment and planning in the energy sector, or, as discretizations of continuous time linear control problems (Dantzig [6]). Often, sparse general large scale linear programs can be permuted to block triangular form.

[1]. This paper is a shortened version of [20]. The research was partially supported by the Office of Naval Research Contract N00014-75-C-0267, the Department of Energy Contract EY-76-S-03-0326 PA#18, and the National Science Foundation Grants MCS76-81259, MCS76-20019, and ENG77-06761 all at Stanford University.

[2]. Current address: IBM T.J. Watson Research Center, Yorktown Heights, New York 10598.

Linear programs of this type are difficult to solve using standard simplex method techniques because they usually require a disproportionately large number of iterations (Beale [1]). As a result, several attempts have been made to devise more efficient algorithms by exploiting this structure.

As far back as 1955 (Dantzig [5]) a key empirical observation was made: <u>for dynamic (time-staged) models similar type activities are likely to persist in the basis for several periods</u>. To take advantage of this possibility an algorithm for block triangular systems was outlined. The central idea was to use a <u>square</u> block triangular "artificial basis", that is, one with square blocks on the diagonal, together with a factor to correct for the off-square diagonal blocks of the true basis. Under persistence the true basis would have diagonal blocks that were nearly square. This would imply that the correction would differ from the identity matrix in only a few columns and so require little work in being updated from one iteration to the next. Further, to solve equations with the square block triangular factor, only the diagonal blocks need be factorized. This could yield a substantial savings in storage requirements, as well as an increase in speed.

In this paper we shall extend these ideas by examining the persistence property in detail, by characterizing factorizations in which the correction factor has the minimal number of structural columns, and by showing that only a small submatrix of the correction factor need be maintained to execute the simplex algorithm. Then we shall show how the factorization can be updated from one iteration to the next. In the remaining sections we shall discuss the implementational details, provide an estimate of the work involved, and present some computational experience.

In [13] Kallio and Porteus also employ a square block triangular factorization along the lines suggested in [4]. However their factorization and methods of updating differ substantially from ours. The factorization we consider always maintains the square block triangular factor as a submatrix of A; in [13], this factor contains some transformed columns, making it denser than ours and more difficult to store, but yielding a sparser correction factor. No computational experience is reported in [13].

Other methods devised specifically for this class of problems have considered mainly staircase structures. See for example the nested decomposition algorithm of Ho and Manne [11] and the block factorization techniques of Loute [15] and Wollmer [25].

Bisschop and Meeraus [3] have proposed a method for very general L.P.'s that is not based on any triangularization or block triangularization techniques, but uses an idea that to some extent underlies the method presented here: If we wish to solve the nonsingular system $Dy = d$, but have at our disposal a nonsingular factorized matrix E, where E differs from D in (say) k columns, then there exists a suitable k x k nonsingular matrix P such that the solution to $Dy = d$ can be obtained by solving systems involving only E and P. The method presented in [3] uses this idea to save substantially on storage costs over the conventional LU methods, by confining the growth of nonzeros to P. Based on the empirical results presented here, our method appears to achieve even greater savings in storage.

## 2. Persistence in Staircase Models

Time-staged staircase linear programs are often discrete versions of an underlying continuous process that evolves as the solution to a differential equation. One such continuous model is the following linear optimal control problem with mixed constraints on the state and control variables:

$$\text{minimize} \quad \int_0^T \{c^T x(t) + d^T u(t)\}dt$$

$$\text{subject to} \quad \left. \begin{array}{l} \dot{x} = Ax + Bu + a \\ 0 = Cx + Du + b \end{array} \right\} \quad \ell \text{ equations}$$

$$x(t) \geq 0 \quad u(t) \geq 0 \quad t \in [0,T] \quad x(0) = p \;.$$

It has been shown in [19] that if $(x(\cdot), u(\cdot))$ is a nondegenerate extreme point solution of the above (convex) constraint set, and if $u(\cdot)$ is piecewise analytic, then the interval [0,T] can be partitioned into a collection of open intervals such that on each interval, precisely $\ell$ components of (x,u) are strictly positive, and the remaining ones are identically zero.

In such cases we would hope that as we make finer and finer the discrete approximation to the continuous-time problem, the basic solutions will tend to behave more and more in this way, i.e., have the same activities basic over several consecutive time periods, and moreover have precisely $\ell$ of them do so. That this is often the case in practice was observed by Dantzig [4].

In the following we shall investigate this persistence property by looking at the structural properties of bases in general staircase models where each step has the same size, say $\ell$ x r. Such models have constraints of the form

$$A_1 \, x_1 = b_1$$

$$B_{t-1} \, x_{t-1} + A_t \, x_t = b_t, \quad t = 2,..., K$$

$$x_t \geq 0$$

where the coefficient matrices $A_t$ and $B_t$ are all $\ell$ x r.

Let B be any basis, and let $P_t$ and $Q_t$ be those submatrices of $A_t$ and $B_t$ respectively that appear in B. $P_t$ and $Q_t$ will have the same number of columns (which can vary from one t to the next); for all t, $P_t$ and $Q_t$ will have $\ell$ rows. For t = 1,..., K let $k_t$ be such that $P_t$ and $Q_t$ are $\ell$ x ($\ell + k_t$). Clearly $k_t$ may be of any sign and is a measure of the "off-squareness" of period t in this basis. Note that since B is square, $\Sigma \, k_t = 0$.

The following theorem gives bounds on how far off-square the basis can be along the diagonal.

2.1. <u>Theorem</u>: For all t and s such that $1 \leq t \leq t + s \leq K$,

$$-\ell \leq \sum_{j=0}^{s} k_{t+j} \leq \ell \,.$$

<u>Proof</u>: Consider the portion of B between t and t + s:

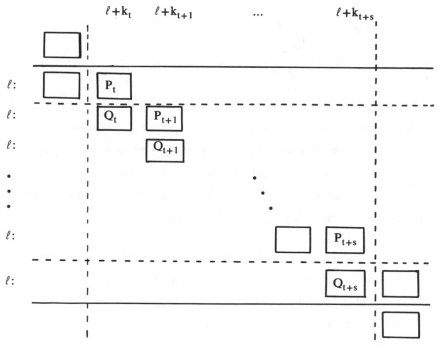

Since B is nonsingular, the columns between the two vertical broken lines are linearly independent. Therefore, since these columns have only zeros outside the two horizontal solid lines, the number of these columns may not exceed the number of rows between the two solid lines, i.e.,

$$\sum_{j=0}^{s} (\ell + k_{t+j}) \leq (s + 2)\ell$$

or

$$\sum_{j=0}^{s} k_{t+j} \leq \ell$$

The nonsingularity of B also implies that the rows between the two horizontal broken lines are linearly independent. A similar argument yields

$$s\ell \le \sum_{j=0}^{s} (\ell + k_{t+j})$$

or

$$-\ell \le \sum_{j=0}^{s} k_{t+j} \; .$$

Combining the above inequalities yields the desired result. □

This result is of interest because it gives the same bound, $\ell$, on the off-square count for any one period as it does on the count for any grouping of consecutive periods. Therefore, in a staircase model where the number of time periods, K, is very large relative to $\ell$, we can choose to group together, say, every s periods to obtain a coarser partition of the matrix in such a way that the ratio of the maximum possible off-square count to the number of periods in any partition, $\ell/s$, is small.

In many economic applications we can attach a stronger although more qualitative significance to the result of Theorem 2.1. Activities like the level of coal production, or the level of an inventory, are usually positive over intervals of time, rather than at points spread haphazardly. In continuous-time this corresponds to a statement about the piecewise continuity of optimal solutions. In discrete-time, we infer that such activities will remain basic over a whole time interval, irrespective of how many time steps constitute the interval. The implication of this for the $\{k_t\}$ is that they will be constant over intervals of time no matter how refined the grid size. Setting $k_{t+j} = k$ for $j = 0, 1,..., s$ in the theorem and noting that k is an integer, we obtain $k = 0$ for grid sizes refined enough that $s \ge \ell$. Thus in solutions where the activities persist in the basis over intervals of time, and each such interval contains at least $\ell + 1$ time steps, the number of activities in each time step is precisely $\ell$. Note, though, that this will not hold true at the end points of the time intervals.

In time-staged models where the matrix structure is block triangular but not staircase, the above discussion and Theorem 2.1 do not apply. However, in cases where the non-staircase part of the block triangular structure contains just a sprinkling of nonzeros, and similar type activities can appear from one period to the next, it seems reasonable to expect a similar behavior.

## 3. The Square Block Triangular Factorization of the Basis

Let B denote the basis matrix, say m x m. B will have a block triangular structure

$$
B = \begin{bmatrix}
B_{11} & & & \\
B_{21} & B_{22} & & \\
& \bullet & & \\
& \bullet & & \\
B_{K1} & B_{K2} \bullet & \bullet B_{KK}
\end{bmatrix}
$$

We shall factorize B into the product of two m x m nonsingular matrices[1]

$$
B = \bar{B} \, F \tag{1}
$$

where -

$$
\bar{B} = \begin{bmatrix}
\bar{B}_{11} & & & \\
\bar{B}_{21} & \bar{B}_{22} & & \\
& \bullet & & \\
& \bullet & & \\
\bar{B}_{K1} & \bar{B}_{K2} \bullet & \bullet \bar{B}_{KK}
\end{bmatrix}
$$

with the diagonal blocks $\bar{B}_{tt}$ square and nonsingular. Note that if B and $\bar{B}$ have k columns in common, then F will contain precisely k unit columns. Thus by suitably permuting the rows and columns of F, we can obtain

$$
F = P \begin{pmatrix} G & \\ H & I \end{pmatrix} Q \tag{2}
$$

where P and Q are permutation matrices, G is (m - k) x (m - k) and nonsingular, and I is the identity matrix of order k.

---

[1] Dantzig [4] works with the (mathematically) equivalent factorization $\bar{B} = BE$ where E is $F^{-1}$.

## 4. Minimal Factorizations

As we shall emphasize later the usefulness of this factorization approach will depend critically on the number of columns that B and $\overline{B}$ have in common, or, more directly, the size of the matrix G in (2). It will be important to know how to obtain a G that is as small as possible.

4.1 Definition: Given a basis B, a square block triangular factorization of B, $B = \overline{B}F$, will be said to be minimal if for any other such factorization, $B = \overline{B}'F'$, $\overline{B}$ has at least as many columns in common with B as does $\overline{B}'$.

4.2 Notation: For two matrices C and D with the same number of rows, let $C \cap D$ denote the submatrix of columns that are in both C and D.

Thus a factorization $\overline{B}F$ is minimal if (number of columns of $B \cap \overline{B}$) $\geq$ (number of columns of $B \cap \overline{B}'$) for all admissible $\overline{B}'$.

We have the following useful characterization of minimal factorizations.

4.3. Theorem: Let B be a given basis. Then a square block triangular factorization of B, $B = \overline{B}F$, is minimal if and only if for t = 1,..., K, $B_{tt} \cap \overline{B}_{tt}$ is a basis for the space spanned by the columns of $B_{tt}$.

Proof: Suppose there is a t such that $B_{tt} \cap \overline{B}_{tt}$ is not a basis for the column space of $B_{tt}$. Then there is a column b of B,

$$b = (O \ b_t \ ... \ b_K)^1$$

such that the augmented matrix

$$W = [B_{tt} \cap \overline{B}_{tt} \ b_t]$$

has full column rank.

---

[1] All partitioned vectors will be written as row vectors but assumed to be column vectors.

Since $\bar{B}_{tt}$ is nonsingular, $b_t$ has a representation $\bar{B}_{tt} \, y = b_t$. Since W has full column rank there is a column $\bar{b}$ of $\bar{B}$,

$$\bar{b} = (O \quad \bar{b}_t \, ... \, \bar{b}_K)$$

such that $\bar{b}_t$ is a column of $\bar{B}_{tt}$ but not of $B_{tt}$, and such that the coefficient of $\bar{b}_t$ the y is nonzero. It follows that the matrix $\bar{B}_{tt}'$ arrived at by interchanging columns $\bar{b}_t$ and $b_t$ is nonsingular. Since $\bar{b}_t$ is not in $B_{tt}$, $\bar{b}$ is not in B. Let $\bar{B}'$ be $\bar{B}$ with the columns b and $\bar{b}$ interchanged. Then $\bar{B}'$ is again square block triangular with nonsingular diagonal blocks. Further, $\bar{B}'$ contains one more column of B than did $\bar{B}$. Therefore the factorization $\bar{B}F$ is not minimal.

Conversely if the factorization $\bar{B}F$ is not minimal there is a factorization $B\,'F\,'$ such that $B \cap B\,'$ has more columns than $B \cap \bar{B}$. Thus for some t, $B_{tt} \cap \bar{B}_{tt}'$ has more columns than $B_{tt} \cap \bar{B}_{tt}$. Also, since $\bar{B}_{tt}'$ is nonsingular $B_{tt} \cap \bar{B}_{tt}'$ has full column rank. Hence $B_{tt} \cap \bar{B}_{tt}$ cannot be a basis for the column space of $B_{tt}$.

This completes the proof. □

4.4. <u>Applications</u> <u>of</u> <u>Theorem</u> <u>4.3</u>: Theorem 4.3 serves two useful purposes. Firstly, it provides us with a constructive method for computing a minimal $\bar{B}$ and F for any B: For each t, find a linearly independent subset, $M_{tt}$, of columns of $B_{tt}$ that is as large as possible. Augment $M_{tt}$ with any other convenient columns, e.g., unit columns, to form a square and nonsingular matrix $[M_{tt} : N_{tt}]$, say. Then let $\bar{B}$ consist of the columns that correspond to the $M_{tt}$, together with the $N_{tt}$, suitably padded with zeros. $\bar{B}$ will thus have $[M_{tt} : N_{tt}]$ as its diagonal blocks. The Theorem tells us that this is a minimal $\bar{B}$.

Secondly, from the proof of the Theorem, we have a means of constructing a minimal factorization from any given square block triangular factorization, $B = \bar{B}F$, as follows. Suppose that the $j^{th}$ column of B, say $b_j$, is not in $\bar{B}$. For some t we can write

$$b_j = (O \quad b_t \, ... \, b_K)$$

The $j^{th}$ column of F, say $f_j$, is the representation of $b_j$ in terms of $\bar{B}$. Since $b_j$ has zeros in the first t - 1 periods, it follows by square block triangularity of $\bar{B}$ that $f_j$ has the form

$$f_j = (O \quad f_t \; ... \; f_K)$$

Thus

$$b_t = \bar{B}_{tt} \, f_t \;.$$

and if $f_t$ has a nonzero component corresponding to any column that is in $\bar{B}$ but not in B, we can replace this column by $b_j$, and so obtain a "better" factorization. Repeating this must yield a minimal factorization.

## 5. Solving Linear Equations

Two major steps in the revised simplex method (Dantzig[5]) are to solve the following equations:

(i) By = a, where a is the entering column and y is used in the minimum ratio test to determine the leaving column;

(ii) $B^T\pi$ = c, where c is an appropriate vector defined at the end of this section and $\pi$ is the vector of prices used to determine the next entering column.

In using the factorization B = $\bar{B}$F to solve, for example, the system By = a, we would respectively solve the systems $\bar{B}z$ = a and Fy = z. The system $\bar{B}z$ = a is easy to solve since $\bar{B}$ is square block triangular, and since $\bar{B}$ (as it will turn out) may always be a submatrix of the problem's coefficient matrix, so inheriting its sparsity properties.

The system Fy = z, however, is more troublesome. From (2) we must solve

$$P \begin{pmatrix} G & \\ H & I \end{pmatrix} Qy = z \;.$$

Both G and H are computed matrices, and are typically between 20% and 40% dense. Further, if G is k x k, then H is (m - k) x k, and for m much larger then k, the storage requirements for H and the work involved in updating H would become prohibitive. In the following we shall show however that it is possible to solve the system $By = a$ without knowledge of H.

To simplify matters we shall assume that we have suitably reordered the columns of B to be $BQ^T$ and the columns of $\bar{B}$ to be $\bar{B}P$, and refer to these reordered matrices as B and $\bar{B}$, so that

$$B = \bar{B} \begin{pmatrix} G \\ H \; I \end{pmatrix}.$$

(For now we are disregarding the original block triangular structure we had B and $\bar{B}$.) Partitioning B and $\bar{B}$ according to the partition of F, we may write $B = (B^1 \; B^2)$ and $\bar{B} = (\bar{B}^1 \; \bar{B}^2)$ where $B^2 = \bar{B}^2 = B \cap \bar{B}$. Likewise, we partition $y = (y^1 \; y^2)$ and $z = (z^1 \; z^2)$. Thus

$$\begin{pmatrix} G \\ H \; I \end{pmatrix} \begin{pmatrix} y^1 \\ y^2 \end{pmatrix} = \begin{pmatrix} z^1 \\ z^2 \end{pmatrix}$$

from which we see that $y^1$ satisfies $Gy^1 = z^1$. Further, since $By = a$ iff $\bar{B} (0 \; y^2) = a - B^1 y^1$, $y^2$ may now be found by solving a second system in terms of $\bar{B}$.

In short, to solve $By = a$, we solve three sets of equations

$$\left. \begin{array}{l} \bar{B}z = a \\ Gy^1 = z^1 \\ \bar{B}w = a - B^1 y^1 \end{array} \right\} \tag{3}$$

where $w = (0 \; y^2)$.

There is a similar easily verified procedure for the system $B^T \pi = c$, again with three sets of equations (upon partitioning $c = (c^1 \; c^2)$).

$$\begin{aligned}
\bar{B}^T \lambda &= (0 \ c^2) \\
G^T \mu &= c^1 - (B^1)^T \lambda \\
\bar{B}^T \pi &= (\mu \ c^2)
\end{aligned} \right\} \tag{4}$$

This shows that we now need only be able to solve systems involving $\bar{B}$ and $G$. Since the quantities $B^1 y^1$ and $(B^1)^T \lambda$ above may be determined at little extra cost, the storage and running costs of the simplex method are now localized to the factorizations of the $K$ diagonal blocks of $\bar{B}$ and the matrix $G$ (that is, apart from the pricing and minimum ratio operations).

The above procedures may be further refined by taking advantage of the nature of the right hand sides. This can be done as follows:

(i) For some t, the incoming column may be written as $a = (0 \ a_t \ \ldots \ a_K)$. Thus the system $\bar{B}z = a$, $\bar{B}$ being square block triangular, may be solved using only the submatrix

$$\begin{bmatrix} \bar{B}_{tt} & & \\ & \cdot & \\ & & \cdot \\ \bar{B}_{Kt} & \cdots & \bar{B}_{KK} \end{bmatrix}.$$

On the average, we would then solve $\frac{1}{2}(K+1)$ sub-systems. Note that this would not apply to the third step in (3) since the right hand side there has no structure in general. Nevertheless the total number of sub-systems to be solved in (3) is now on the average $\frac{3}{2}(K+1)$ for $\bar{B}$ and 1 for $G$.

(ii) During Phase II, the right hand side c is always the same unit vector $e_\ell$ where the objective row appears in row $\ell$ (say) of the coefficient matrix. In modern implementations of the simplex method (Beale [2]) this will not be the case during Phase I since c is set equal to a vector of -1's, 0's and 1's depending on which variables are currently outside their bounds.

For the case $c = e_\ell$ we show that by suitably enlarging $G$, we can solve for $\mu$ in (4) without first computing $\lambda$. Note that since the objective variable associated with the $\ell^{th}$ row of the coefficient matrix is unrestricted in sign, the basis B will always contain its column, which is also the unit vector $e_\ell$. Since we can ensure that $\bar{B}$, too, always contains this column, its corresponding column in F will be a unit vector. Upon suitable rearrangement, we can write

$$F = \begin{pmatrix} G' & \\ H' & I \end{pmatrix}$$

where now

$$G' = \begin{pmatrix} G & \\ h & 1 \end{pmatrix}$$

and h is the corresponding row of H. By partitioning $c = e_\ell$ according to this partition of F, we see that $c^1$ is a unit vector, and $c^2 = 0$. From this it is immediate that $\pi$ may be determined from

$$\left.\begin{array}{l} (G')^T \mu = c^1 \ (= \text{unit vector}) \\ \bar{B}^T \pi = (\mu \ \ 0). \end{array}\right\}$$

Thus, during Phase II, we here too obtain a substantial savings.

## 6. Updating the Factorization

Suppose that we begin an iteration of the simplex method with a basis B and a factorization $\bar{B}F$ of B. At the end of this iteration we obtain a new basis B′ by replacing some column $b_i$ of B with the entering column a and then seek a new factorization $\bar{B}'F'$ of B′. There is in general no unique way of doing this, and determining the best method may require lengthy experimentation. We here present the update that yields a minimal factorization B′ $= \bar{B}'F'$ from a minimal factorization B $= \bar{B}F$. Since its derivation [20] is laborious and lengthy, we shall give only the flow diagram (Figure 1 below).

The two major steps of the update are as follows:

(i) Try to bring the column a into $\bar{B}$ without disturbing square block triangularity. How this is done depends on whether or not a and/or $b_i$ are in $\bar{B}$.

(ii)   Restore the factorization to minimality by seeing if there are any other columns of B, not currently in $\bar{B}$, that can be brought into $\bar{B}$.

The update on $\bar{B}$ can be accomplished by updating at most two of the factorizations of its diagonal blocks, each taking the form of an exchange of columns. The update on G will consist of a possible change in size, up or down by one, and the addition of at most two rank one matrices.

It will be convenient to present the flow diagram in terms of the following fundamental operations:

(i)   INVUPD -- updating the factorization of a $\bar{B}_{tt}$

(ii)   FROW -- solving the system of equations $\bar{B}^T w = e_k$, and then computing $w^T B^1$ to yield the structural part of the $k^{th}$ row of F

(iii)   FCOL -- solving the system of equations $\bar{B}f=b$ to obtain the column f of F corresponding to some column b of B

(iv)   RSWAP -- replacing a row of G

(v)   CSWAP -- replacing a column of G

(vi)   ADD -- increasing the size of G by 1

(vii)   DEL -- decreasing the size of G by 1

(viii)   RANK1 -- performing a general rank one update on G, that is $G' = G+uv^T$ for column vectors u and v

(ix)   FINDB -- seeks a column of B that is not in $\bar{B}$ to replace column $b_i$ in $\bar{B}$

(x)   FIND2 -- seeks two columns, one in B that is not in $\bar{B}$ to replace one in $\bar{B}$ that is not in B.

We remark that the FINDB and FIND2 operations constitute those portions of the flow diagram pertaining to maintaining the minimality of the factorization. Since these operations may be costly, they should only be carried out if it is even costlier to have a G one size larger. Since a minimal factorization is found at every refactorization of B, the optimal policy will be to maintain minimality until the number of iterations remaining before refactorization is below some threshold value. This threshold will depend on the particular implementation as well as the characteristics of the problem being solved. See [20] for a more detailed discussion.

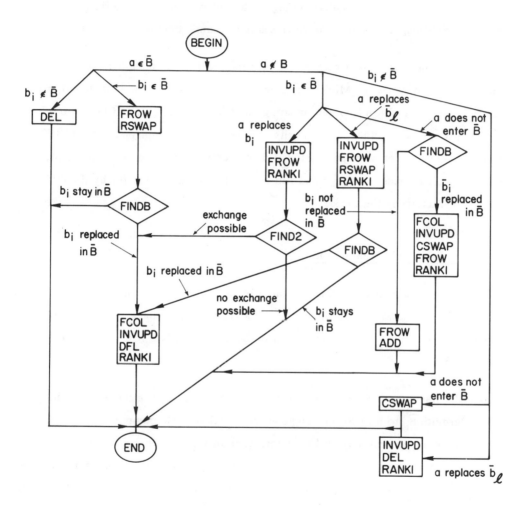

Figure 1: The Update Flow Diagram

## 7. Implementation

The foregoing details of the factorization were implemented in an in-core Fortran program LPBLK on an IBM 370/168. LPBLK is an experimental code, written for the purposes of (i) obtaining an understanding of the behavior of this factorization method, and (ii) establishing its potential for solving large dynamic time period models.

Towards the latter goal we chose another in-core Fortran code MINOS [17,22] as a standard for comparison. MINOS uses an LU factorization of the basis together with Saunders' variant of Bartels-Golub updating, as described in [21], and takes no advantage of any structure other than sparsity. It was easy to implement in LPBLK the CHUZR (choose pivot row) and PRICE (choose entering column) routines of MINOS so that any differences in storage and run time could be accounted for by the different methods of factorizing the basis.

## 7.1. Factorizing and Updating the $\bar{B}_{tt}$

Since the $\bar{B}_{tt}$ are sparse and are updated by an exchange of columns, it was natural to treat them as one would the whole basis of a general sparse linear program. Accordingly we chose to do an LU factorization of $\bar{B}_{tt}$ and to update it using the method of Forrest and Tomlin [8].

The LU factorization was computed as outlined in [23] and restated here briefly:
(i)     Permute $\bar{B}_{tt}$ to near lower triangular form with a single bump.
(ii)    Rearrange so that only an LU factorization of the bump is necessary.
(iii)   Determine the LU factorization of the bump, pivoting for sparsity: First sort the bump columns according to a Markowitz-like criterion. Then pivot in this order, at each stage choosing as pivotal element the one whose row has the lowest count of non-zeros subject to a relative size tolerance.

## 7.2. Factorizing and Updating G

From Section 6, we see that whichever factorization we choose for G, it will have to be such that rank one updates and changes in size can be easily and stably accomplished. Unlike the $\bar{B}_{tt}$, G is a computed matrix so that the need for stable methods is even greater for G. In the light of these considerations, a QR factorization of G seemed to be the logical choice. Since G turns out to be typically between 30% and 50% dense there would be little need for sparsity considerations, thus allowing the implementation to be relatively straightforward.

The factors Q and R are defined to be those orthogonal and upper triangular matrices respectively that satisfy $Q\,G = R$ . One method of computing these factors is to apply a sequence of elementary Householder transformations[1] to G to reduce it to the matrix R. By a suitable column interchange strategy (see [14], [26]) one can ensure that

$$| r_{ii} | \geq | r_{jk} | \quad j,k \geq i, \ \text{all i}$$

where $R = (r_{ij})$. This enables one to use the ratio

$$\rho = \max | r_{ii} | / \min | r_{ii} |$$

as a lower bound on the condition number of G. Thus excessively large values of $\rho$ may be taken as an indication that G is ill-conditioned and that it or the whole factorization should be recomputed perhaps with a different choice of $\bar{B}$. (See Section 8.2).

When G is modified by a matrix of rank one, the factors Q and R may be updated by applying a sequence of plane rotations. These are orthogonal matrices obtained by imbedding a 2x2 orthogonal matrix

$$\begin{pmatrix} c & s \\ s & -c \end{pmatrix}$$

suitably in the identity. In general one may require up to 2(p-1) such rotations for the general rank one update, where G is pxp. See [9] for further details.

---

[1]Householder transformations are orthogonal matrices of the form $P - 2uu^T$ where $u^T u = 1$.

## 7.3. Recomputing the Factorization $B = \bar{B}F$

In Section 4 we sketched a method of finding a minimal factorization for any given B. This involved finding a maximal linearly independent subset of columns of each $B_{tt}$ and then augmenting them appropriately with unit columns to obtain the corresponding $\bar{B}_{tt}$. It turns out that the method of computing the LU factors of a sparse non-singular matrix outlined in Section 7.1 can be easily adapted for this task: Proceed precisely as indicated there but with a rectangular bump instead, and skipping a column when there is no nonzero of some suitable magnitude available as a pivotal element. After the first pass, a second can be made through the set of rejected columns in case some of them are now suitable for pivoting. Finally, insert unit columns in those rows in which no pivoting has yet taken place.

Once we have found our factor $\bar{B}$ we represent, in terms of $\bar{B}$, each column of B that was rejected in the above process. Thus if p columns were rejected, we solve p systems of the type $\bar{B}y = b_i$ and form G by extracting from each y those components corresponding to the augmented unit vectors.

Remarks: (i) It is usually difficult to determine numerically the rank of a matrix, or, more precisely in our context, to pick the threshold for determining whether or not the next column is linearly dependent on the current set. Observe that, in our case, stopping a few columns short of obtaining a maximal independent set is of little consequence since it only implies that we shall be working with a factorization that is slightly off from being minimal. Thus in the interests of stability it would pay us to use stricter tolerances than in the case when we know what the rank of a matrix should be.

(ii) Observe that since the $B_{tt}$ are assumed roughly square, it is still reasonable to preorder the bump columns with the same criterion as before.

## 8. Experimental Results

As our test problems we used 5 dynamic models, described below.

| Name | Rows | Structural Columns | Periods | Non-zeros | % Density | Bounds |
|------|------|------|---------|-----------|-----------|--------|
| PILOT8 | 626 | 1376 | 8 | 6026 | .7 | YES |
| SC205 | 204 | 202 | 19 | 551 | 1.3 | NO |
| SCRS8 | 491 | 1169 | 4 | 4029 | .7 | NO |
| SCSD8 | 398 | 2750 | 39 | 11334 | 1.0 | NO |
| SCTAP1 | 301 | 480 | 10 | 3372 | 2.3 | NO |

Table 1:  Problem Statistics

PILOT8 is an 8 period SIGMA version of the PILOT energy model currently under study at Stanford University.  It has a staircase structure together with a few nonzeros sprinkled in the lower block triangle.  For a description of the model see [7].  The others have staircase structures and were derived from real models in economic planning (SC205), agricultural production scheduling (SCRS8), engineering design (SCSD8) and dynamic traffic control (SCTAP1).  For more detailed descriptions of these models see [12].

The runs were made in several different ways.  For PILOT8 we started from an advanced basis and stopped the run after 2 minutes CPU time.  This yielded approximately 500 iterations.  The same advanced basis was used for all these runs.  For SCSD8, we started from an advanced basis and allowed the runs to terminate at optimality, again requiring about 500 iterations.  The remaining models were all started from an identity basis and allowed to terminate at optimality.  The iteration counts for these were 230 for SC205, 870 for SCRS8, and 400 for SCTAP1.  Duplicate runs were made at varying degrees of partitioning, e.g., combining every 3 periods to form new periods each 3 times as large but 1/3 as many in number.  In addition were recorded runs of PILOT8.S, the PILOT8 model rescaled using the geometric mean [24].

## 8.1 Estimated breakdown of the work per iteration

Through a careful analysis of the update operations and the implementation as described above, it was possible to estimate the total number of multiplications per simplex iteration aside from the pricing and minimum ratio steps. These estimates can be extremely valuable in assessing the relative efficiency of alternative methods without them first being implemented. In [20] the estimates are derived in detail; here we present only the main results.

The estimates were made in terms of the following parameters which would be problem dependent.

K:  Number of time periods.

$\eta$:  Average number of nonzeros in the L and U factors of a typical $\overline{B}_{tt}$ (assumed independent of t).

p:  Average size of G.

q:  The frequency with which G and the QR factorization are recomputed.

$\xi$:  Average number of times per iteration that a row or column of G is generated by solving a system in terms of $\overline{B}$.

$\mu$:  Average number of systems per iteration to be solved in terms of a single $\overline{B}_{tt}$ other than for $\xi$ above.

$\gamma$:  Average number of systems per iteration to be solved in terms of Q.

$\alpha$:  Average number of sweeps[1] per iteration.

$\rho$:  Average proportion of plane rotations applied per sweep.

Very surprisingly, the last five parameters were observed to be more or less problem independent, taking on the following values:

$$\xi \approx 1, \quad \mu \approx 2, \quad \alpha \approx 2.5, \quad \gamma \approx 3.2, \quad \rho \approx .6.$$

The above value for $\gamma$ is for the case when Q is maintained explicitly, i.e., the Householder transformations and plane rotations are accumulated into a single matrix. An alternative method would be to keep Q in product form, in which case generating a

---

[1]A "sweep" is the application of a sequence of plane rotations to reduce a given vector to a multiple of a unit vector.

column of Q requires the solution of a system of equations. It then turns out that $\gamma \simeq 3.7$.

With these values of $\xi$, $\mu$, $\alpha$, $\gamma$ and $\rho$, the following table resulted.

| | | |
|---|---|---|
| $\bar{B}$ | Phase I | $(4K + 3)\,\eta$ |
| | Phase II | $(3K + 3)\,\eta$ |
| G | Explicit | $\dfrac{p}{q}\left\{ \dfrac{1}{2}\eta(K+1) + \dfrac{4}{3}p^2 \right\} + 13.2p^2$ |
| | Product | $\dfrac{p}{q}\left\{ \dfrac{1}{2}\eta(K+1) + \dfrac{2}{3}p^2 \right\} + 11.1\,pq + 9.7\,p^2$ |

Table 2: Estimated multiplications per iteration

Table 3 below gives the estimated number of multiplications for each of the test runs. Based on run times and numerical stability, a good rule of thumb for q with the explicit form appears $q = 30$. In the product form there is a trade off between saving time in not multiplying each rotation into Q, and in accumulating a great number of factors that slow down the equation solution times. $q_m$ is then the theoretically optimum time at which to recompute and refactorize G.

From this table, we see that with the exception of the PILOT8 run with $K=8$, the explicit form for Q is always superior. However it should be noted that there will be models for which the product form is much superior. These will be ones having high values of p, i.e., G very large, and high values of $\rho$, the i.e., G very dense. Also in cases where the storage of Q explicitly requires much core, the product form too will have an advantage.

As far as the degree of partitioning is concerned, notice that refining the partition has the effect of increasing the size of G (necessarily) and decreasing the sizes of the diagonal blocks of $\bar{B}$. Thus we would expect some intermediate partition to be best. However, we see that on PILOT8, PILOT8.S and SCSD8 it is far better to have a small number of periods than a large number. The main reason for this is that a disproportionate amount of computation goes into updating G, so that an increase in the size of

G has a far greater effect than a corresponding increase in the number of rows of a period. This makes it clear that any improvement in the methods of representing G will be very worthwhile.

| Name | K | $\bar{B}$ (1000's) | | $q_m$ | G (1000's) | | Total |
|------|---|------|------|-------|----------|---------|-------|
| | | Phase | | | Explicit | Product | Phase II |
| | | I | II | | q = 30 | q = $q_m$ | Explicit |
| PILOT8 | 8 | 13.9 | 10.5 | 23 | 81.3 | 77.6 | 91.8 |
| | 4 | 20.5 | 16.0 | 17 | 39.7 | 40.3 | 55.7 |
| PILOT8.S | 8 | 14.6 | 11.1 | 23 | 67.8 | 69.2 | 78.9 |
| | 4 | 22.7 | 18.1 | 17 | 31.4 | 33.8 | 49.5 |
| SC205 | 19 | 1.2 | .9 | 3 | .3 | .6 | 1.2 |
| SCRS8 | 4 | 8.8 | 6.8 | 12 | .3 | .8 | 7.1 |
| SCSD8 | 39 | 4.1 | 3.1 | 12 | 32.4 | 34.7 | 35.5 |
| | 20 | 6.8 | 5.1 | 12 | 14.9 | 17.7 | 20.0 |
| | 13 | 8.5 | 6.5 | 11 | 7.7 | 10.7 | 14.2 |
| | 8 | 10.1 | 7.7 | 12 | 4.4 | 6.5 | 12.1 |
| | 5 | 12.5 | 9.8 | 14 | 1.0 | 2.1 | 10.8 |
| SCTAP1 | 10 | 1.9 | 1.5 | 6 | .4 | .9 | 1.9 |
| | 5 | 2.9 | 2.3 | 7 | .4 | .8 | 2.7 |

Table 3 : Estimated multiplications per iteration

Finally we remark that the number of multiplications has turned out to be a good indicator of the work involved based on a near straight line relationship with the cpu times in Table 4.

## 8.2. A Comparison with MINOS

We ran each of the problems on MINOS, adjusting tolerances and refactorization frequencies where necessary to obtain run times that were as fast as possible. The runs on LPBLK were all made with Q stored explicitly, and maintaining the minimality of G at each step. Table 4 below resulted.

ETA is the number of nonzeros in the representation of each $\bar{B}_{tt}$ for LPBLK and the whole basis for MINOS. The times are given in CPU milliseconds. All figures were computed as averages over the whole run.

| Name | K | ETA($\eta$) | LPBLK | | MINOS | | |
| | | | Size of G (p) | Time/ Iteration | ETA | Spikes | Time/ Iteration |
|---|---|---|---|---|---|---|---|
| PILOT8 | 8 | 423 | 71 | 302 | 14582 | 152 | 203 |
| | 4 | 1117 | 46 | 240 | | | |
| PILOT8.S | 8 | 433 | 66 | 288 | 15815 | 130 | 188 |
| | 4 | 1168 | 42 | 235 | | | |
| SC205 | 19 | 15 | 4 | 30 | 1021 | 23 | 25 |
| SCRS8 | 4 | 502 | 4 | 92 | 3335 | 38 | 65 |
| SCSD8 | 39 | 26 | 44 | 213 | 4358 | 56 | 130 |
| | 20 | 82 | 31 | 177 | | | |
| | 13 | 131 | 21 | 163 | | | |
| | 8 | 295 | 17 | 163 | | | |
| | 5 | 545 | 7 | 158 | | | |
| SCTAP1 | 10 | 47 | 6 | 55 | 2204 | 28 | 40 |
| | 5 | 130 | 5 | 56 | | | |

Table 4

From the CPU times in the above table, we see that MINOS is approximately 22% faster than LPBLK on PILOT8, PILOT8.S, SC205 and SCSD8. It is much faster on SCRS8 and SCTAP1.

Scaling seems to have had the same effect on both codes. In the case of MINOS the scaled columns would make the pivot elements of triangle columns more acceptable (in a relative test) and so force fewer triangle columns to become spikes. It would appear that a similar argument could be made for LPBLK: if more pivot elements become acceptable then the FINDB and FIND2 operations would be more successful in keeping the size of G to a minimum.

On the stability side, the only models to present problems were PILOT8 and PILOT8.S. In the runs recorded here, LPBLK experienced no trouble; however there were some advanced bases for PILOT8 that LPBLK failed to refactorize while MINOS succeeded. As mentioned in Section 7.2 the ratio

$$\max |r_{ii}| / \min |r_{ii}|$$

is a lower bound on the condition number of G. On PILOT8 and PILOT8.S this ratio was typically of the order $10^8$, sometimes going as high as $10^{10}$. With the other models, this ratio was at most $10^3$. The PILOT model has always been badly conditioned and a code like MPSIII [16] often needs to force unit columns into the basis and lose feasibility in order to maintain stability. MINOS is perhaps the only code that has not suffered unduly on PILOT.

One apparent cause of PILOT's illconditioning is the presence of dense 12x12 Leontief matrices imbedded in each period. The variables corresponding to these Leontief systems were all free (no upper or lower bounds)[1] and hence always in the basis. In the case of LPBLK, by the manner in which the $\overline{B}_{tt}$ are defined during a complete refactorization (see Section 7.3), the dense Leontief columns would be considered last for introduction into $\overline{B}$ and many of them would not "make it". This

---

[1]This was done since it was possible to demonstrate in advance that in any optimal solution, these activities would always be at a positive level.

would result in several columns of G being transforms of these Leontief columns, a fact that seemed intuitively wrong. Accordingly we forced all the Leontief columns into $\bar{B}$, with the result being a much improved factorization: the lower bound on the condition number of G dropped by as much as two orders of magnitude.

## 8.3. Storage Considerations

Table 5 below contains statistics on the average number of nonzeros required to represent the basis factorization. The first two columns give the average number of nonzeros (for both LPBLK and MINOS) taken over the whole run.[1] The third column gives the average number of nonzeros recorded immediately after a MINOS refactorization. This table shows that the basis storage requirements of LPBLK are between 23% and 60% of those of MINOS, a substantial savings.

The method of Bisschop and Meeraus [3] leaves the initial factorization of the basis untouched and instead updates a certain augmented matrix of size kxk where k is the number of columns in which the initial and current bases differ. A lower bound on their storage requirements is therefore given by the number of nonzeros in the initial representation. Assuming that their initial factorization is an LU as performed by MINOS we may use the third column of Table 9 as the lower bound. Comparing this with the column for LPBLK, we observe, at the appropriate degree of partitioning that LPBLK is far superior on all except SCRS8 and SCSD8 where it is no worse than the lower bound.

---

[1]These were taken directly from Table 4, with the entries in the LPBLK column being set equal to $K \eta + 3/2 \, p^2$ for the explicit form of Q.

| Name | K | Average Over The Run | | Average at refactorization (MINOS) |
|------|---|------|------|------|
| | | LPBLK | MINOS | |
| PILOT8 | 8 | 10946 | 14582 | 11874 |
| | 4 | 7642 | | |
| PILOT8.S | 8 | 9998 | 15815 | 12088 |
| | 4 | 7318 | | |
| SC205 | 19 | 309 | 1021 | 641 |
| SCRS8 | 4 | 2032 | 3335 | 2040 |
| SCSD8 | 39 | 3918 | 4358 | 2331 |
| | 20 | 3082 | | |
| | 13 | 2365 | | |
| | 8 | 2794 | | |
| | 5 | 2798 | | |
| SCTAP1 | 10 | 524 | 2204 | 1609 |
| | 5 | 688 | | |

Table 5: Storage Requirements

## 9. Conclusion

As matters stand, the computational results of the previous section show that LPBLK saves considerably on storage requirements. However, from the run times it is apparent that LPBLK comes close to, but is not yet competitive with the best available software. It seems that problems on which it will excel will be ones like PILOT8 with more periods and perhaps a greater proportion of nonzeros in the lower block triangle.

However, the preceding analysis clearly shows that cheaper ways have to be found to deal with G. Any method that can halve the work involved with G will make a 30% difference in the run time, so making it very competitive on the test problems presented. One possibility currently under study is to use the factorization $Q\,G^T = R$ but then discarding Q altogether and instead maintaining G explicitly. To solve the equations $Gx = g$ we would first solve two triangular systems $R^TRy = g$, and then set $x = G^Ty$. Since G is typically between 30% and 50% dense it may well be possible to store it as a sparse matrix, so that forming $G^Ty$ will be inexpensive. Thus solving equations this way will involve no more work than when we keep Q. Where we save considerably is in the update, since now we need only apply the rotations to R.

This approach was first used by Gill and Murray [10] in the context of the simplex method for dense systems. Paige [18] has shown that the above method of solving the equations is stable, involving only $K(G)$ (instead of $K^2(G)$) in the error bound. However, to solve for the price vector, we need to solve equations of the form $G^Tx = g$ (refer to Section 5). This requires the solution of the equations $R^TRx = Gg$ where now the term $K^2(G)$ does enter into the error bound. Since the price vector only indicates which column to bring into the basis, and is not used in any further computation, less stability here can be tolerated. Further, in Phase II we can simplify the procedure since we always solve $G^Tx = e_r$ where $e_r$ is a fixed unit vector. Multiplying this relation on the left by Q yields $Rx = Qe_r$. Therefore by maintaining the $r^{th}$ column of Q explicitly from one iteration to the next, we can solve these equations by means of a single triangular system. This is both stable and inexpensive.

In addition to the choice of representation for G, the fine tuning aspects involving the FINDB and FIND2 operations alluded to in Section 6 have yet to be properly implemented and tested. Some experimental runs made along these lines have been very encouraging.

More work needs also to be done on the stability side in general. There is a definite need for more care in deciding which columns go into $\bar{B}$ and which are transformed to become columns of G.

## Acknowledgments

The authors are grateful to Michael Saunders, John Tomlin and Margaret Wright for the many hours of discussion and deliberation, and also for the use of codes they had written. Without them this work would not have been possible.

Thanks are also due to James Ho for supplying the SC205, SCRS8, SCSD8 and SCTAP1 test problems.

The authors wish to express special thanks to Michael Saunders for his helpful suggestions during the preparation of this manuscript.

## References

[1]   E.M.L. BEALE, "Sparseness in Linear Programming" in Large Sparse Sets of Linear Equations, J.K. Reid (ed.) (1971), Academic Press London, pp. 1-15.

[2]   E.M.L. BEALE, "Advanced Algorithmic Features for General Mathematical Programming Systems," in Integer and Nonlinear Programming, J. Abadie (ed.) (1971), North-Holland Publishing Company, London.

[3]   J. BISSCHOP and A. MEERAUS, "Matrix Augmentation and Partitioning in the Updating of the Basis Inverse," Mathematical Programming, 13, 3 (1977), 241-254.

[4]   G.B. DANTZIG, "Upper Bounds, Secondary Constraints, and Block Triangularity in Linear Programming" Econometrica, 23, April 1955, pp. 174-183.

[5]   G.B. DANTZIG, Linear Programming and Extensions, (1963), Princeton University Press, Princeton, New Jersey.

[6]   G.B. DANTZIG, "Large-Scale Systems Optimization with Application to Energy", Technical Report SOL 77-3, April 1977, Department of Operations Research, Stanford University, Stanford, California.

[7]   G.B. DANTZIG and S.C. PARIKH, "At the Interface of Modeling and Algorithms Research" Technical Report SOL 77-29, October 1977, Department of Operations Research, Stanford, University, Stanford, California.

[8]   J.J.H. FORREST and J.A. TOMLIN, "Updating Triangular Factors of the Basis in the Product Form Simplex Method", Mathematical Programming, 2 (1972), 263-278.

[9]   P.E. GILL, G.H. GOLUB, W. MURRAY and M.A. SAUNDERS, "Methods for Modifying Matrix Factorizations", Math Comp. 28, p.p. 505-535 (1974).

[10]  P.E. GILL and W. MURRAY, "A Numerically Stable Form of the Simplex Algorithm" Linear Algebra and its Applications, 7, (1973), 99-138.

[11]  J.K. HO and A.S. MANNE, "Nested Decomposition for Dynamic Models", Mathematical Programming, 6 (1974) pp. 121-140.

[12]  J.K. HO, "Implementation and Application of a Nested Decomposition Algorithm", Proceedings of the Bicentennial Conference on Mathematical Programming, November-December 1976, Gaithersburg, Maryland.

[13]  M. KALLIO and E.L. PORTEUS, "Triangular Factorization and Generalized Upper Bounding Techniques", Operations Research, 25, 1, (1977) 89-99.

[14]  C.L. LAWSON and R.J. HANSON, Solving Least Squares Problems, Prentice-Hall, New Jersey, (1974).

[15]  E. LOUTE, "A Revised Simplex Method for Block Structured Linear Programs", Doctoral Dissertation, Applied Sciences Faculty, Catholic University of Louvain, Belgium, 1976.

[16]  MPS III Users Manual (1973), Management Science Systems, Rockville, MD.

[17]  B.A. MURTAGH and M.A. SAUNDERS, "A Large-Scale Nonlinear Programming System (for Problems with Linear Constraints) -- User's Guide", Technical Report SOL 77-9, February 1977, Department of Operations Research, Stanford University, Stanford, California.

[18]  C. PAIGE, "An Error Analysis of a Method for Solving Matrix Equations", Math. Comp., 27, 122, April 1973, 355-359.

[19]  A.F. PEROLD, "Fundamentals of a Continuous Time Simplex Method", Ph.D. Thesis (November 1978), Department of Operations Research, Stanford University, Stanford, California.

[20]  A.F. PEROLD and G.B. DANTZIG, "A Basis Factorization Method for Block Triangular Linear Programs", Technical Report SOL 78-7, April 1978, Department of Operations Research, Stanford University, Stanford, California.

[21] M.A. SAUNDERS, "A Fast, Stable Implementation of the Simplex Method Using Bartels-Golub Updating", in Sparse Matrix Computations, ed., J.R. Bunch and D.J. Rose, Academic Press, New York, New York, (1976), pp. 213-226.

[22] M.A. SAUNDERS, "MINOS System Manual", Technical Report SOL 77-31, December 1977, Department of Operations Research, Stanford University, Stanford, California.

[23] J.A. TOMLIN, "Pivoting for Size and Sparsity in Linear Programming Inversion Routines", J. Inst. Maths. Applic. (1972), 10, 289-295.

[24] J.A. TOMLIN, "On Scaling Linear Programming Problems", Mathematical Programming Study 4, (1975), 146-166.

[25] R.D. WOLLMER, "A Substitute Inverse for the Basis of a Staircase Structure Linear Program", Mathematics of Operations Research, 2, 3, (1977), 230-239.

[26] M.H. WRIGHT and S.C. GLASSMAN, "Fortran Subroutines to Solve the Linear Least-Squares Problem and Compute the Complete Orthogonal Factorization", Technical Report SOL 78-8, April 1978, Department of Operations Research, Stanford University, Stanford, California.

# On Combining
# the Schemes of Reid and Saunders
# for Sparse LP Bases

## David M. Gay*

Abstract.  Both M. A. Saunders and J. K. Reid have
proposed efficient schemes for maintaining a factored
representation of the sparse basis matrices that arise in
large-scale linear programming (LP).  Saunders's scheme
lends itself particularly well to an "out-of-core" implementation, while Reid's scheme can reduce the growth of
the factors it maintains by means of clever permutations.
Both schemes efficiently implement the Bartels-Golub method
for updating LP bases.  Following a suggestion of D.
Goldfarb, we investigate combining these schemes, with the
aim of discovering whether this might lead to more efficient
basis handling codes, particularly out-of-core codes or
in-core codes run on paged machines.  We also investigate
use of an idea attributed to Beale for exploiting block
structure that is in the basis at "reinversion" time.  This
sometimes allows one of the basis factors in Saunders's
scheme to be represented more compactly.  Our numerical
experience with a basis handling code incorporating the
above ideas shows that they can be quite worthwhile.  This
experience also suggests that the Hellerman-Rarick $P^4$ pivot
order is better suited to Saunders's scheme than the block
Markowitz order and that, barring excessive paging, Reid's
scheme alone may yield the best in-core algorithm.

1.  Introduction.  Both Saunders [Sau76] and Reid [Rei75]

have proposed efficient schemes for maintaining a factored

representation of the sparse basis matrices that arise in

large-scale linear programming (LP).  Saunders's scheme

lends itself particularly well to an "out-of-core" imple-

mentation, while Reid's scheme can reduce the growth of the

factors it maintains by means of clever permutations.  Both

schemes efficiently implement the Bartels-Golub method for

*MIT/CCREMS, Cambridge, MA 02139.

updating LP bases [Bar71]. Following a suggestion of
Goldfarb [Gol75], we investigate combining these schemes,
with the aim of discovering whether this might lead to more
efficient basis handling codes, particularly out-of-core
codes or in-core codes run on paged machines. We simultaneously investigate a scheme for exploiting block structure
that may be present in the basis when it is initially
factored or refactored ("reinverted").

The next section sketches Saunders's scheme, while §3
considers two potential improvements in this scheme: use
of Reid's scheme to handle a relatively small but crucial
submatrix and use of an alternate representation of certain
matrix factors. In §4 we describe an implementation that
incorporates these schemes into an experimental modular
linear programming package called XMP. We report on
computational experience in §5 and offer conclusions in §6.

   2. Saunders's Scheme. In Saunders's scheme, the (suitably permuted) basis B is factored as B = LU. Initially
(after each reinversion) L is a lower triangular matrix
and U is upper triangular with ones on the diagonal.
When B is updated (i.e., when a column of B is replaced
by another column), the corresponding column of U is
replaced and elementary permutations and row operations are
applied to U to reduce it to upper triangular form. To
the matrix L, which is maintained in product form, are
adjoined the inverses of these elementary operations, so
that the factorization B = LU is maintained. The factors

which constitute L are always accessed sequentially and
hence may be kept conveniently in secondary storage until
needed. It is necessary to maintain U in a more accessi-
ble form. Saunders chooses the initial ordering of B by
the Hellerman-Rarick $P^3$ or $P^4$ strategy [HelR71, HelR72],
which results in a U having only a small number of
columns with nonzeros above the diagonal ("spikes").
Thus, for an appropriate permutation matrix P,
$P^T UP = \begin{bmatrix} I & R \\ 0 & F \end{bmatrix}$ , where F is upper triangular. When B
is updated, R may lose a row and gain a column or simply
have one column replaced by another. Otherwise $P \begin{bmatrix} R \\ 0 \end{bmatrix}$
remains unchanged and hence R (or $P \begin{bmatrix} R \\ 0 \end{bmatrix}$) may be kept
in secondary storage until needed. It is only necessary to
store F in main memory; since F normally is of relatively low order, this usually presents no difficulties.
Often F contains enough zeros that it is worthwhile to
store it as a sparse matrix.

   3. Improvements. We now consider two strategies that
may reduce the number of nonzeros in the factors stored
by Saunders's basic scheme. The first applies to the
initial factorization of B: when computing the initial
L and U (or reinverting), it may prove worthwhile to
take advantage of the block structure of B, which is available as a byproduct of the $P^4$ ordering scheme. We may do
this by adapting to Saunders's scheme an idea which
Orchard-Hays [Orc68, p.82] attributes to Beale. Suppose
B has the block lower triangular form

$$(3.1) \qquad B = \begin{bmatrix} B_{11} & & & \\ B_{21} & B_{22} & & \\ \vdots & \vdots & \ddots & \\ B_{k1} & B_{k2} & \cdots & B_{kk} \end{bmatrix}$$

and suppose $B_{jj} = L_j U_j$, $1 \leq j \leq k$, where $L_j$ is lower triangular and $U_j$ is upper triangular. Then

$$L = \begin{bmatrix} L_1 & & & \\ B_{21}U_1^{-1} & L_2 & & \\ B_{31}U_1^{-1} & B_{32}U_2^{-1} & \ddots & \\ \vdots & \vdots & \ddots & \\ B_{k1}U_1^{-1} & B_{k2}U_2^{-1} & \cdots & L_k \end{bmatrix} \quad \text{and} \quad U = \begin{bmatrix} U_1 & & & \\ & U_2 & & \\ & & \ddots & \\ & & & U_k \end{bmatrix}.$$

Rather than explicitly computing the portions of $L$ below the diagonal blocks, i.e., $B_{ij}U_j^{-1}$ for $i > j$, and thereby suffering fill-in, it often saves storage (for $j < k$) to retain $L_j$, $U_j$, and $B_{ij}$ for $i > j$. To solve linear systems involving the original $L$, it suffices to access these submatrices sequentially, so this more compact representation of $L$ may also be kept conveniently in secondary storage.

How worthwhile the above representation of $L$ is depends partly on how $L$ is stored. For example, if the entries of the constraint matrix are stored as single-precision numbers and computed quantities are stored in double precision, or if the constraint matrix entries are stored supersparsely [Ka171], then greater savings are possible than may be obtained when all data are stored explicitly in the same precision. In some cases, retaining $U_j$ and $B_{ij}$ may actually require more storage than $B_{ij}U_j^{-1}$ would take, and in other cases the increase in overhead caused by dealing separately with $U_j$ and $B_{ij}$ may outweigh the storage savings. For the implementation described in the next section, however, our computational experience suggests that it is often worthwhile to exploit block structure in this way.

The second strategy for reducing the number of nonzeroes in the basis factors deals with updating these factors. Reid [Rei75] has shown that it is sometimes possible to reduce the number of elementary elimination matrices adjoined to L by appropriately permuting U. When applying Reid's ideas to Saunders's scheme, it suffices to permute F, which then becomes the product of a permutation matrix, an upper triangular matrix, and another permutation matrix. The computational experience which we report in §5 suggests that Reid's permutations often lead to some storage savings when used with Saunders's scheme.

4. An Implementation under XMP. To see how worthwhile the above ideas might be in practice, we have incorporated them into a basis handling code designed to be used with XMP, a modular linear programming package [Mar78]. The code keeps all data in main memory. As much as possible, it seeks to access storage sequentially, with the aim of reducing paging on large problems. Thus, for example, the nonzeroes of a column of $P\begin{pmatrix} R \\ 0 \end{pmatrix}$ are stored consecutively,

and the columns of $P\begin{pmatrix} R \\ 0 \end{pmatrix}$ are accessed in order of their appearance in memory. Moreover, the nonzeroes in the columns of the initial L matrix are also stored consecutively, even though the nonspike columns of L are also available in the original constraint matrix.

Factorization of the basis involves two phases: first a symbolic phase selects a tentative pivot order based on the sparsity pattern of the basis, then a numerical phase computes the actual basis factors, modifying the tentative pivot order if necessary to avoid small pivots and exploiting block structure when this appears worthwhile. Once the basis has been factored, the F matrix is accessed and updated by means of slightly modified versions of Reid's subroutines LA05B and LA05C [Rei76], so that we may benefit from his clever updating technique. The following paragraphs discuss these points in more detail.

We have experimented with two ways of choosing the tentative pivot order used in factoring the basis: the $P^4$ method of Hellerman and Rarick [HelR71, HelR72] and a block Markowitz scheme. In both cases it is first necessary to find the block structure of the basis. We do this with the help of Harwell subroutines MC21A [Duf77] and MC13D [DufR78], thus following the approach recommended by Gustavson [Gus76]. To compute the $P^4$ pivot order, we apply the $P^3$ algorithm [HelR71] to each nontrivial diagonal block (using our own $P^3$ implementation). Similarly, to compute the block Markowitz pivot order, we find the Markowitz

pivot order [Mark57] for each nontrivial diagonal block
(by passing the transpose of the current diagonal block to
a version of Reid's LA05A that has been modified to deal
with the sparsity pattern of the basis).

One byproduct of the symbolic phase is a list of spike
columns, i.e., columns of the basis that have at least one
nonzero above the main diagonal of the basis when the basis
is permuted according to the tentative pivot order. The
second phase of our basis factoring code uses this spike
list along with the tentative pivot order as it computes
the numerical factors. Columns of the basis are processed
according to their appearance in the tentative pivot order.
Nonspike columns with acceptable pivots are copied directly
into the representation for L, while spike columns must
first undergo some elementary transformations. The tentative pivot order is accepted unless an unacceptably small
pivot is encountered, one which would yield a column of L
whose diagonal entry would have magnitude no more than,
say, PIVTOL times that of some entry beneath it. In this
case we use the scheme that Saunders [Sau76] suggests: scan
the spikes remaining in the current block in search of one
which would have a larger pivot element; if one is found,
then interchange it with the current column. Otherwise
retain the current column (and its small pivot), unless
this would result in a zero pivot. If it would, then
replace the current column by an artificial one, i.e., a
column having a one in the pivot row and zeros elsewhere

and whose solution component is required to be zero. In
the testing reported below, we always used PIVTOL = 0.1.
This worked well on all problems considered in §5 except
problem Stair, which sometimes had an excessive number of
column interchanges. (Saunders's test results [Sau76]
suggest that PIVTOL = 0.01 would have been a better choice
for problem Stair.)

After each block of columns is processed, the potential
savings from exploiting block structure as in §3 are
assessed. If the savings appear worthwhile for block j,
then the $L_j$ and $U_j$ of §3 are recovered from the columns
of L just stored, and the blocks $B_{ij}$ (for i > j) are
extracted from the original constraint matrix.

The matrix F and the transformations adjoined to L
when the basis is updated are stored in the format used
by Reid's LA05 routines [Rei76], with minor changes which
make it easy to increase the dimension of F. One change
is the introduction of an overdimension parameter. The
other is the separation of Reid's permutation arrays from
his other integer arrays. This separation enables us to
avoid shifting entries in the permutation arrays when the
dimension of F increases.

In effect, we represent the permuted basis B as the
product $B = LP \begin{pmatrix} I & R \\ O & F \end{pmatrix} P^T$, where L is not changed by updates and F is a general sparse matrix that starts out
as an upper triangular matrix when the basis is factored,
but loses this form during basis updates. Since

$B^{-1} = P \begin{pmatrix} I & -R \\ 0 & I \end{pmatrix} \begin{pmatrix} I & 0 \\ 0 & F \end{pmatrix}^{-1} P^T L^{-1}$, we may readily reduce the problem of operating with $B$ (i.e., of performing FTRANs and BTRANs) to the problem of solving systems of linear equations involving $F$. An appropriately modified version of Reid's LA05B solves this reduced problem for us.

When the basis is updated, either $R$ and $F$ have one column exchanged for another, or else $R$ loses a row and $F$ and $R$ gain a column (and the permutation $P$ is updated accordingly). In this second case, we reduce the problem of updating $F$ to that of changing a column of $F$ as follows: we adjoin as a new first row of $F$ the row removed from $R$ (with a one in the (1,1) position) and as a new first column the first standard unit vector, i.e., $(1,0,0,\ldots,0)^T$. It then suffices to exchange this new first column for the one that actually belongs there. In both cases of updating $F$ we are thus able to employ an appropriately modified version of Reid's LA05C to finish the update and give us the benefit of Reid's updating method.

In our implementation, storage for $L$ and $R$ occupies one area and storage for $F$ another. $L$ occupies the top of its storage area, $R$ the bottom. At reinversion time, the percentage of available storage allotted to $F$ and $L/R$ is based on how much storage was in active use by these matrices just prior to the reinversion. Should either the $F$ or the $L/R$ area overflow during factorization, then the share of storage allotted to these areas is changed to agree with the ratio of nonzeros currently stored in these areas; if this change is sufficiently large (and in the same direction as any previous change made during the current factorization), then the factorization is attempted again; otherwise storage overflow is recognized.

When  R  is updated, new entries are stored after all current entries if space permits, and discarded entries continue temporarily to occupy storage.  If space does not permit updating in this way, then storage occupied by the dicarded entries is recovered.  If this still does not allow  R  to be updated, then the basis must be refactored.

Refactorization because of storage overflows occurred very infrequently in our testing, and the initial storage allocations to  F  and  L/R  made at factorization time had very rarely to be modified.

5.  Computational Experience.  We have tested the basis handling code described above on five medium-scale linear programming problems.  Some statistics on these problems and the basis factorizations computed in our tests are shown in Table I.  Each time that the basis was factored anew we computed the basis factors in three ways: by calling Reid's LA05A to see how the Markowitz pivot order (MK) would fare, and by using our basis factoring code with the $P^4$ and block Markowitz (BM) tentative pivot orders.  The factorization from the BM order was retained and updated during subsequent iterations until the basis was again

factored anew.  Table I gives the minimum and maximum
value of certain statistics over all the factorizations
performed as well as the average value of these (and other)
statistics.  Notes following Table I explain the various
statistics in more detail.

| I T E M | See Note | Stair | Shell | Powell | Seba | Ganges |
|---|---|---|---|---|---|---|
| Number of Rows | 1 | 357 | 537 | 549 | 516 | 1311 |
| Number of Columns | 2 | 824 | 2312 | 1403 | 1551 | 2992 |
| Nonzeroes in Constraint Matrix | 2 | 4226 | 5437 | 7254 | 5775 | 8441 |
| % Density of Constraint Matrix | 2 | 1.43 | 0.44 | 0.94 | 0.72 | 0.22 |
| Number of Iterations | 3 | 1693 | 1182 | 4570 | 1361 | 1638 |
| Basis Factorizations | 4 | 17 | 18 | 65 | 18 | 25 |
| N O N Z E R O E S in Newly Computed Basis Factors........ | | | | | | |
| Min. % Saved by Blocks: $P^4$ | 5 | -0.3 | 0 | 0 | -16.9 | 0 |
| Max. " " " " " | 5 | 49.5 | 0 | 55.7 | 0 | 0 |
| Mean " " " " " | 5 | 14.9 | 0 | 11.2 | -5.9 | 0 |
| Min. " " " " BM | 5 | -0.3 | 0 | 0 | -23.4 | 0 |
| Max. " " " " " | 5 | 72.2 | 0 | 72.1 | 8.0 | 2.6 |
| Mean " " " " " | 5 | 19.2 | 0 | 15.0 | -5.1 | 0.1 |
| Min. % Increase $P^4$ over BM | 6 | -38.7 | 0 | -6.5 | -23.0 | -1.1 |
| Max. " " " " " | 6 | 39.1 | 0 | 8.6 | 35.2 | 1.2 |
| Mean " " " " " | 6 | 3.6 | 0 | 1.3 | 4.4 | 0.03 |
| Min. " " BM " MK | 7 | 0 | 0 | -0.5 | -22.2 | 0 |
| Max. " " " " " | 7 | 179.8 | 0 | 28.5 | 9.8 | 3.6 |
| Mean " " " " " | 7 | 24.0 | 0 | 13.0 | -2.4 | 1.3 |
| Mean % in Initial L: LA05A | 8 | 19.0 | 0 | 16.0 | 0.3 | 4.9 |
| " " " " ": BM | 8 | 89.4 | 100.0 | 90.1 | 99.8 | 97.7 |
| F: Mean % Increase BM/$P^4$ | 9 | 35.8 | 0 | 70.6 | 32.9 | 16.9 |
| F: Overall % Increase BM/$P^4$ | 9 | 68.6 | 0 | 59.5 | 48.5 | 26.2 |
| Mean % Blocks Exploited: $P^4$ | 10 | 40.1 | 0 | 95.6 | 51.9 | 4.0 |
| " " " " : BM | 10 | 39.0 | 0 | 93.2 | 43.5 | 8.0 |
| Mean Initial % F Density: $P^4$ | 11 | 24.7 | 0 | 42.3 | 33.0 | 45.1 |
| " " " " " : BM | 11 | 20.1 | 0 | 34.2 | 33.6 | 42.4 |
| Mean Final % F Density: BM | 11 | 33.6 | 13.8 | 37.7 | 18.5 | 15.2 |
| Mean % Init. Spike Inc. BM/$P^4$ | 12 | 28.7 | 0 | 43.0 | 14.7 | 13.7 |

Table I: Problem and Factorization Statistics

Notes on Table I

1.  The constraint matrix includes a free row for the objective function (and sometimes free rows for other functionals of interest).  The row count listed includes free rows.

2.  The XMP modules used in our tests add one slack, surplus, or artificial variable for each row in the constraint matrix.  Thus a diagonal matrix of ones and minus ones is appended to the end of the original constraint matrix.  The column count includes these appended columns, as do the figures for the number of nonzeros and density of the constraint matrix.

3.  The XMP modules used start with an all-logical initial basis (i.e., the adjoined diagonal matrix of ones and minus ones mentioned in Note 2).  This probably results in substantially more iterations than if a good "crash" heuristic were used to select the initial basis.

4.  The number of basis factorizations listed includes one for the initial basis and one for the final basis: when optimality (or infeasibility) is detected, the basis is refactored and the optimality conditions rechecked.

5.  "% Saved by Blocks" refers to the percentage by which the number of nonzeros in the representation of the L  matrix computed at each factorization would increase if block structure were not exploited as described in §3.  The test that decided when to exploit block structure considered the storage required for the (single- and double-

precision) values of the nonzeros in the representation of
L rather than the actual number of these nonzeros.
Because of this, there were times when the alternate representation was used for some columns of L even though it
resulted in more nonzeros. This explains the negative
entries for the Seba problem.

6. "% Increase $P^4$ over BM" refers to the percentage
increase in the number NZP4 of nonzeros in the basis
factors computed from the $P^4$ tentative pivot order beyond
the similar number NZBM of nonzeros from the block Markowitz order, i.e., 100% x (NZP4 - NZBM)/NZBM.

7. "% Increase BM over MK" means
100% x (NZBM - NZMK)/NZMK, where NZBM is as in Note 6 and
NZMK is the number of nonzeros in the basis factors computed by Reid's LA05A using the Markowitz pivot order (as
modified by threshold pivoting -- see [Rei76]).

8. "Mean % in Initial L" means the average percentage
of significant nonzeros in the basis factors L and U of
the factorization B = LU that are in the L matrix,
i.e., 100% x LNZ/(LNZ + UNZ), where LNZ and UNZ are
the number of nonpivot nonzeros in L and U respectively. L and U are permuted triangular matrices. The
nonzeros of the permuted diagonals of L and U are the
"pivot nonzeros" and are excluded from LNZ and UNZ,
since they are insignificant so far as fill-in (in solutions to linear systems involving L and U) is concerned.
The average is taken over all basis factorizations except

that of the initial all-logical basis.

9. "F: Mean % Increase BM/P$^4$" means the average percentage increase in the number of nonzeros FBM in the F matrix computed from the BM pivot order beyond the number FP4 computed from the P$^4$ order, i.e.,

100% x (FBM - FP4)/FP4, averaged over all factorizations. "F: Overall % Increase BM/P$^4$" means the similar percentage increase in the sum SFBM of nonzeros in F in all the basis factorizations using the BM order beyond the sum SFP4 corresponding to the P$^4$ order, i.e.,

100% x (SFBM - SFP4)/SFP4.

10. "%Blocks Exploited" means the percentage of non-singleton blocks $B_{jj}$ in (3.1) for which it appeared worthwhile to use the alternate representation of §3 for the corresponding block of columns of L.

11. "Initial % F Density" means the percentage of non-zeros in the upper triangular portion of the F matrix produced when the basis was factored. "Final % F Density" means the similar percentage of nonzeros in F just before the basis was refactored, i.e., after all the times it was updated.

12. "% Init. Spike Inc. BM/P4" means the increase in the number of initial spikes produced from the block Markowitz tentative pivot order above the number from the P$^4$ order, i.e., 100% x (NSBM - NSP4)/NSP4, where NSBM is the number of columns in the F and R computed using the BM order and NSP4 is the similar number using the P$^4$ order.

Discussion of Table I. The behavior summarized in Table
I varies widely from problem to problem. This is probably
due in part to the unusual features that some of the problems have. Virtually all the bases considered in solving
the Shell problem, for example, could be permuted to lower
triangular form. The Seba and Ganges problems both seem
in large part to have the structure of network problems.
Problem Stair's constraint matrix has a staircase form.
Only the Powell problem has no immediately obvious special
structure.

In general, the block Markowitz pivot order resulted in
slightly fewer nonzeros on average than did the $P^4$ order.
(It also caused the factorizations to take longer, on average, to compute: 25% longer for the Powell problem, 16%
longer for the Stair problem, and about 5% longer for the
others.) Exploiting block structure gave greater savings,
on average, with the block Markowitz order than with the $P^4$
order, but if block structure had not been exploited at all,
then both orders would have averaged nearly the same number
of nonzeros in the basis factors. With regard to the number
of spikes (i.e., of columns in F and R) as well as the
number of nonzeros in F, $P^4$ clearly had the advantage over
the BM order. The number of nonzeros in F is of particular interest, since they are the only nonzeros in the basis
factors that are not accessed sequentially and could not be
so easily kept in secondary storage. All things considered,
the $P^4$ order seems to enjoy a modest advantage over the BM

order for use in Saunders's basis handling scheme.

Exploiting block structure proved quite worthwhile for the Stair and Powell problems. Even so, Reid's LA05A was able to produce factorizations with still fewer nonzeros (by an average of 24% and 13% respectively on these problems) by using the full Markowitz pivot order (with threshold pivoting). The LA05A factorization has a further advantage in that a much smaller percentage of its fill-in causing nonzeros are in the L matrix. When the LU factorization produced by LA05A is updated, the new column brought into U is therefore likely to have significantly fewer nonzeros than are the corresponding columns brought into the F and R of Saunders's scheme. Reid's scheme therefore is likely to enjoy a slower growth rate of nonzeros during updating (at the cost of having to access the rows of the U matrix randomly). His scheme therefore appears decidedly advantageous to the modified Saunders scheme for problems of such size that all data can be kept in main memory, at least so long as excessive paging does not occur. Of course, Saunders's scheme is primarily intended for use on problems so large that some data must be kept in secondary storage. Since it accesses most of its data sequentially, it might also prove to be a better choice than Reid's scheme on problems where there is sufficient virtual memory to use his LA05 routines, but where excessive paging occurs when they are used. Limits of time and budget prevented us from seeing how our in-core implementation of the modified Saunders scheme compares with LA05 on problems where significant paging occurs.

Updates. Table II gives some statistics on the updates that our test code performed on the basis factors initially computed using the block Markowitz pivot order. The "New Spikes" figures give the fraction of updates in which F and R gained a new column. Though the average number of nonzeros gained per update by the basis factors is relatively small, the "Min." and "Max." figures show that the actual number for a particular update can be quite large. These figures also show that F and R can lose a considerable number of nonzeros when one of their dense columns is replaced by a sparse one. Figure 1 illustrates how the numbers of nonzeros in L, F, and R vary during a typical sequence of updates. Contrary to the discussion near the end of §4, in Table II and Figure 1 we regard the updates as adding to L the transformations necessary to bring F back to permuted upper triangular form.

The last row in Table II gives the percentage of updates in which no transformations had to be added to L. Reid's updating technique [Rei75] certainly makes this percentage higher than it otherwise would have been, so it is clearly worthwhile to use Reid's updating scheme on the F matrix (though our test results do not illustrate precisely how worthwhile this is).

| I T E M | P R O B L E M | | | | |
| --- | --- | --- | --- | --- | --- |
| | Stair | Shell | Powell | Seba | Ganges |
| New Spikes per Update | 0.45 | 0.96 | 0.40 | 0.75 | 0.90 |
| New Nonzeros per Update | | | | | |
| L: Min. | 0 | 0 | 0 | 0 | 0 |
| L: Max. | 120 | 2 | 51 | 32 | 16 |
| L: Mean | 9.5 | 0.0 | 7.7 | 2.6 | 0.8 |
| R: Min. | -182 | -22 | -257 | -387 | -188 |
| R: Max. | 202 | 36 | 307 | 401 | 234 |
| R: Mean | 32.9 | 7.3 | 51.4 | 20.1 | 22.1 |
| F: Min. | -99 | -3 | -53 | -30 | -18 |
| F: Max. | 119 | 17 | 83 | 38 | 30 |
| F: Mean | 11.1 | 1.7 | 8.1 | 3.2 | 2.9 |
| % of Updates with No Change to L | 39.8 | 99.9 | 30.4 | 61.9 | 84.6 |

Table II:  Updating Statistics

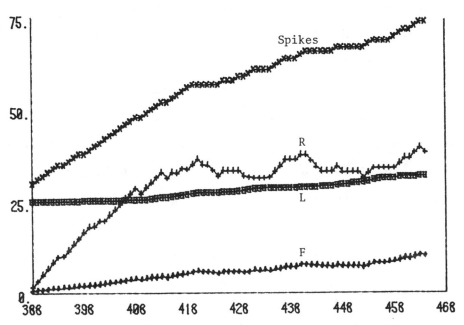

Figure 1:  Spikes and 0.01 × nonzeros in L, R, and F
after updates 388 through 464 on problem Stair.

6. Conclusions. We have studied two potential improvements in the LP basis handling scheme of Saunders [Sau76]. Both aim to reduce the storage required for the basis factors. The first applies when the basis is refactored. It sometimes produces a significantly more compact representation of one of the basis factors by using an alternate representation for some blocks of its columns. The second potential improvement, use of Reid's updating technique [Rei75] on a critical submatrix, aims to reduce the number of nonzeros added to the basis factors when the basis is updated. Our test results show that both can be worthwhile.

In light of the good performance that the Markowitz pivot order often enjoys [DufR74], we thought that a block Markowitz pivot order might work well with Saunders's scheme. Our test results, however, suggest that the Hellerman-Rarick $P^4$ pivot order [HelR71, HelR72] is better suited to this scheme.

For LP problems small enough that the basis factors can be kept in main memory, our test results and those of [Rei75] suggest that Reid's LA05 routines [Rei76] offer the best currently available way to handle LP bases, at least so long as excessive paging does not occur. On problems where LA05 suffers excessive paging, it appears likely that the modified Saunders scheme would prove worthwhile, since it accesses most of its data sequentially. However, it remains to be seen whether this is actually so.

The modified Saunders scheme looks quite promising for

use on problems so large that some of the basis factor data
must be kept in secondary storage. The evidence presented
in [Sau72, Fig. 3] and [Sau76] suggests that Saunders's
scheme enjoys a much smaller nonzero growth rate than does
the standard product form of the inverse. The modified
Saunders scheme can only fare still better in this comparison. How it fares on sparsity grounds against the Forrest-
Tomlin scheme [ForT72] is less clear. Reid's evidence
[Rei75] shows that the permutations chosen by his updating
technique can lead to a substantially lower nonzero growth
rate than is enjoyed by the Forrest-Tomlin scheme. However, the L matrix used by Saunders's scheme in the basis
factorization B = LU is denser than the one used in
Reid's scheme, and this will cause a somewhat higher nonzero growth rate during updating. Although it is therefore
unclear how it compares with the Forrest-Tomlin scheme as
regards nonzero growth during updating, the modified
Saunders scheme has a distinct advantage in terms of
numerical stability, since it includes pivoting for
stability during updating. This should increase the
number of updates that can be performed before numerical
inaccuracy makes it necessary to refactor the basis. All
told, the modified Saunders scheme holds definite promise
and deserves further study.

Acknowledgements. I thank J.K. Reid and E. Yip for
supplying problems Powell, Stair, and Shell, and W.
Northup for supplying problems Seba and Ganges. I also

thank W. Northup for many useful discussions and for his considerable help with the computational testing. This research was supported in part by NSF Grants MCS 76-01311 and MCS 76-01311 A01.

## REFERENCES

[Bar71]    R. H. BARTELS, _A stabilization of the simplex method_, Numer. Math. 16 (1971), pp. 414-434.

[Duf77]    I. S. DUFF, _On algorithms for obtaining a maximum transversal_, Report CSS.49, Computer Science and Systems Division, A.E.R.E. Harwell, Oxon., England, 1977.

[DufR74]   I. S. DUFF and J. K. REID, _A comparison of sparsity orderings for obtaining a pivotal sequence in Gaussian elimination_, J. Inst. Math. Appl. 14 (1974), pp. 281-291.

[DufR78]   I. S. DUFF and J. K. REID, _An Implementation of Tarjan's algorithm for the block triangularization of a matrix_, ACM Trans. Math. Software 4 (1978), pp. 137-147.

[ForT72]   J. J. H. FORREST and J. A. TOMLIN, _Updating triangular factors of the basis to maintain sparsity in the product form simplex method_, Math. Programming 2 (1972), pp. 263-278.

[Gol75]    D. GOLDFARB, comment in discussion of Saunders's paper at the Symposium on Sparse Matrix Computations, Argonne National Laboratory, 1975.

[Gus76]    F. GUSTAVSON, _Finding the block lower triangular form of a sparse matrix_, pp. 275-289 of _Sparse Matrix Computations_, edited by J. R. Bunch and D. J. Rose, Academic Press, New York, 1976.

[HelR71]   E. HELLERMAN and D. C. RARICK, _Reinversion with the preassigned pivot procedure_, Math. Programming 1 (1971), pp. 195-216.

[HelR72]   E. HELLERMAN and D. C. RARICK, _The partitioned preassigned pivot procedure_ $(P^4)$, pp. 67-76 of _Sparse Matrices and their applications_, edited by D. J. Rose and R. A. Willoughby, Plenum Press, New York, 1972.

[Kal71]    J. E. KALAN, _Aspects of large-scale in-core linear programming_, Proc. ACM Annual Conf. (1971), pp. 304-313.

[Mark57]   H. M. MARKOWITZ, <u>The elimination form of the in-verse and its application to linear programming</u>, Management Sci. 3 (1957), pp. 255-269.

[Mar78]   R. MARSTEN, <u>XMP: a structured library of subrou-tines for experimental mathematical programming</u>, Technical Report No. 351, Management Information Systems Dept., Univ. of Arizona, Tucson, AZ 85721, 1978.

[Orc68]   W. ORCHARD-HAYS, <u>Advanced linear-programming com-puting techniques</u>, McGraw-Hill, New York, 1968.

[Rei75]   J. K. REID, <u>A sparsity-exploiting variant of the Bartels-Golub decomposition for linear program-ming</u>, Report CSS 20, Computer Science and Systems Division, A.E.R.E. Harwell, Oxon., England, 1975.

[Rei76]   J. K. REID, <u>FORTRAN subroutines for handling sparse linear programming bases</u>, Report AERE - R8269, A.E.R.E. Harwell, Oxon., England, 1976.

[Sau72]   M. A. SAUNDERS, <u>Large-scale linear programming using the Cholesky factorization</u>, Report STAN-CS-72-252, Computer Science Dept., Stanford Univ., Stanford, CA 94305, 1972.

[Sau76]   M. A. SAUNDERS, <u>A fast, stable implementation of the simplex method using Bartels-Golub updating</u>, pp. 213-226 of <u>Sparse Matrix Computations</u>, edited by J. R. Bunch and D. J. Rose, Academic Press, New York, 1976.